GREAT
VICTORIAN
INVENTIONS

GREAT VICTORIAN INVENTIONS

*Novel Contrivances and
Industrial Revolutions*

Caroline Rochford

AMBERLEY

ACKNOWLEDGEMENTS

A brief note of thanks to my parents, Ruth and Christopher, for their support, and also to the team at Amberley for overseeing the project. Finally, an extra special 'thank you' to my fantastic husband, Michael, without whose boundless knowledge and unceasing assistance this book would have been finished months ago.

First published 2014
Amberley Publishing
The Hill, Stroud
Gloucestershire, GL5 4EP

www.amberley-books.com

Copyright © Caroline Rochford, 2014

The right of Caroline Rochford to be identified as the Author
of this work has been asserted in accordance with the
Copyrights, Designs and Patents Act 1988.

British Library Cataloguing in Publication Data.
A catalogue record for this book is available from the British Library.

ISBN 978 1 4456 3617 7 (paperback)
978 1 4456 3645 0 (ebook)

Typeset in 10pt on 12pt Celeste Pro.
Typesetting and Origination by Amberley Publishing.
Printed in the UK.

CONTENTS

Preface   6

1   The Home   7

2   Leisure   44

3   Fashion, Art and Design   68

4   Education, Work and Industry   84

5   Electrical Innovations   122

6   Lights, Camera, Action!   142

7   Health and Safety, Victorian Style   170

8   Science and Nature   210

9   Technology and Communication   232

10 Travel and Transport   250

Index of Inventions   287

PREFACE

Have you ever heard of the Victorian Whispering Machine? This discreet earpiece promised to narrate the novels of the day into the ears of astounded listeners, and was conceived more than a century before the iPod came into existence. Can you imagine a bicycle railway that ran across America, or a firm of high street opticians that sold spectacles to short-sighted horses? What about the domestic hand grenades that were thrown into the blazes of raging house fires in order to extinguish the flames?

Far-fetched as they may sound, these and countless other contrivances were actually designed by inventors during the final decades of Queen Victoria's reign.

The Industrial Revolution had kick-started the development of the modern world, in which people were encouraged to better themselves through hard work and a good education, and newfangled technology meant that ideas and innovations began to spread far across the nation like the winding tracks of George Stephenson's railway. Consequently an array of radical designs and novel inventions sprang from the fertile minds of these pioneering Victorians, and the late nineteenth century gave rise to the likes of Thomas Edison's electric light, Louis Le Prince's amazing moving photographs, and Alexander Graham Bell's telephone. These were the most revolutionary developments of the era, but what about the long-forgotten propositions and fanciful theories that didn't quite make the grade?

Taken from editions of Cassell's *Family Magazine*, published in 1884, 1885, 1887, 1892 and 1893, this book documents contemporary reports from Britain and around the globe, which describe the latest cutting-edge technology and scientific breakthroughs of the day. From the precursors of modern technology to the most implausible and downright dangerous devices, this volume offers an honest, entertaining and enlightening glimpse at the way in which our forebears used to think, live and dream.

1

THE HOME

The Moveable House (1884)

The late nineteenth century was a time of great prosperity. Following the Industrial Revolution, Victorian towns and cities across Great Britain had given rise to the imposing mills and factories that dominated the crowded urban landscapes. These burgeoning businesses had opened their doors to thousands of jobseekers, who flocked in their droves in search of work. Over time vast suburbs were built on the outskirts of settlements, furnishing workers and their families with places to live. This in turn led to the development of public transport, meaning

that families could reside in better accommodation, far from the smoky slums that were quickly emerging around the inner cities, and the wealthier could afford to seek out even more desirable residences in quieter locations.

In 1884 a French architect named Monsieur Poitrineau exploited the aspirations of the middle classes by devising a number of useful buildings that were, in effect, moveable houses. These portable dwellings were said to be particularly useful for tourists, artists, sportsmen, and others who were 'roughing it' for a season in country parts, as well as those who simply fancied a change of scenery. The homes were made in flat-packed pieces which could be easily put together by the family at any desired location, and, moreover, the entire building could be shifted about by the aid of a low horse-drawn waggon.

To move the structure, the waggon was backed underneath the raised floor of the building and fixed in place; the legs of the building were shortened by unscrewing the lower part, and the whole edifice then rested on the waggon, allowing it to be moved to a new location with the greatest of ease. The boards around the foot of the building hung down when the house was in place in order to hide the legs, and the interiors were fitted up with all the necessities of modern life.

The Portable Greenhouse (1885)

The novel idea of mobile buildings quickly captured public imagination, and by 1885 it was possible for families to take their greenhouse with them when moving home. The flat-packed greenhouse could be easily dismantled and removed along with one's other household belongings and then reconstructed in a new garden. It was designed and built in a convenient size, and constructed of seasoned wood, glazed with sheet glass, and fitted with gutters and pipes to carry rainwater away.

The Glass House (1893)

An artistic housing development was made in Chicago, when seventeen immovable homes made entirely from glass bricks were erected. For the sake of lightness the bricks were hollow, and for artistic purposes they were dyed and moulded to form intricate patterns on the surface, illuminating the interiors with a brilliant spectrum of light.

It was a Glaswegian named FitzPatrick who invented the process whereby these state-of-the-art bricks were cast. Though there were some initial public misgivings, he reassured the glass-house purchasers that these iridescent buildings were capable of resisting atmospheric deterioration much better than ordinary brick, and, moreover, they could be easily cleaned.

The Straw Villa (1887)

Another curious type of Victorian home was exhibited at a spectacular star-studded event held at Earl's Court, London, in May 1887. Amid the colourful displays of American ingenuity, evening bands, and dazzling Wild West shows – that were said to have even impressed Queen Victoria – one particular exhibit at the American Exhibition stood out from the rest. Though the old fable has long warned against building a house from straw, visitors to the attraction were amazed to find a new type of villa that had been constructed entirely from this unconventional material. The straw, of ordinary kind, had been compressed into an artificial wood, which was then used to construct every conceivable part of the house, from the foundations, floors and walls all the way up to the roof and chimneys. The two-storey villa was said to be of artistic design, and, rather astonishingly, it was claimed to be both fireproof and waterproof, though any suggestion of it also being wolf-proof was entirely speculative.

Cooking with Gas (1887)

While movable homes and eco-friendly structures were a fascinating concept for the Victorian homeowner, it was generally felt, as far as the discerning housewife was concerned, that the most interesting inventions were those which led to improvements in the kitchen. By 1887 a London-based engineer had hit upon an idea that helped to revolutionise the domestic economy. This man had turned his attention to the as yet unrealised idea of cooking with gas, and exhibited a new type of oven, which he called the 'Parisian roaster'. The meat was to be cooked inside an enamelled chamber protected by a glass door, through which the chef could gauge the condition of the joint. No gas-mixed air could come into contact with the food, thereby eliminating any risk of contamination, and it was even possible to keep the temperature constant – a task which had hitherto proved somewhat difficult owing to the reliance upon an open fire to cook the food. With this new oven, the radiant heat from the gas burners did the cooking, and direct contact between the flames and the food was avoided.

The Reflecting Oven (1893)

During the latter part of the century significant advances in the gas pipe network brought about the availability of early gas ovens to the general public. Similarly, heating and cooking by means of electricity was also making rapid strides in both Britain and America. Houses were already being fitted with the electric current for lighting purposes, and by the 1890s electric frying and stew pans, kettles, flat irons, curling tongs, and other domestic utensils were in daily use in these modernised households.

However, there were many in society who did not favour such drastic changes, and countless more in poorer parts of the country who simply could not afford the luxury of gas and electricity. Those living in rural areas relied almost exclusively upon oil lamps and candles to generate light. Such families would have no doubt found the novel new oven illustrated in

the above image advantageous. The joint of meat, which was hung at right angles to the front of a traditional open fire, was not directly roasted by the latter, but by reflection of the radiant heat from the two tin reflectors, R R. The food was thus equally cooked all round and required no turning, and therefore very little exertion was required on the part of the chef.

The Oven Heat Indicator (1887)

For conventional stoves and ovens a useful heat indicator was developed, as shown in the accompanying engraving; this was particularly beneficial for ovens without glass doors. The indicator resembled a watch dial, and was helpfully marked 'Cold', 'Warm', 'Bread', 'Meat', 'Pastry' and 'BURNING!', while it also indicated the temperature in Fahrenheit up to 600°. The dial was attached to the front of the door, so that it became unnecessary for the cook to keep opening and closing it to tell how hot the oven was, or if, indeed, the pastry was on fire.

The Indicating Cooking Skewer (1893)

A most serviceable device, which indicated when food was thoroughly cooked through, is shown opposite. It consisted of an ordinary skewer, which enclosed a column of mercury, and was surmounted by an indicator with a brass front. On the brass was marked the point which the mercury should touch once the meat had reached the optimum temperature, even in the middle, which, if incorrectly cooked, often came out cold. The skewer was pushed into the centre of the dish in question directly after it had been taken from the oven, and in around forty-five seconds the mercury would rise to the given mark, providing the food had been adequately cooked. So delicate was this test that if one side of the joint was not quite so well done as the other, the column of

mercury would appear appreciably lower when the skewer was thrust into the former than when placed in the latter.

The Automatic Egg Cooker (1892)

A dish which was considered much easier to cook was the humble boiled egg, but even this could prove problematic if the timing was not precise, leading to many a spoilt breakfast. That was until Parisian inventor, Monsieur Mesdran, decided an *oeuf* was simply an *oeuf*, and he came up with his 'automatic egg cooker'.

The hollow central cylinder, A, shown overleaf, slid over a second cylinder, B, containing a notch in which the trigger, C, would catch when the upper cylinder was pressed down, carrying with it the tray upon which the eggs stood. This action was carried out inside the saucepan, shown in the lower diagram. The eggs were just covered by cold water, and the spirit lamp underneath was lit. As soon as the eggs were done the pressure of the steam released the trigger, and a coiled spring within the

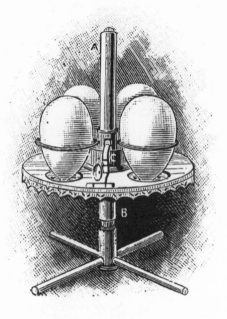

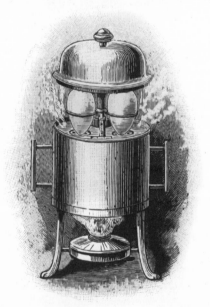

central cylinder mechanically lifted the eggs out of the water, as shown. For still greater convenience, the lid of the saucepan was automatically raised at the same time.

The Egg-Boiling Spoon (1893)

An alternative method for boiling eggs involved the use of a special spoon, which had a sandglass embedded in its stem. While the eggs were boiling the spoon was allowed to stand in the saucepan, as shown, in order to allow the sand to run down, timing the cooking to perfection.

The bowl of the spoon was perforated with holes to allow the hot water to run out once the eggs had been scooped from the pan, thus depositing them in a dry state on the breakfast table.

The Infusing Teapot (1892)

The perfect accompaniment to one's boiled egg was, of course, a refreshing cup of tea – a beverage favoured by individuals from

every class. However, in Victorian times the concept of the teabag had yet to be realised (the first having been hand-sewn in around 1903), so to have tea without the annoyance of loose tealeaves was a widely felt desire. In ordinary teapots the leaves, which were allowed to circulate freely in the hot water, were filtered at the last moment through a strainer, which seldom prevented all the leaves from passing into the cup. To remedy this, a new leafless vessel, dubbed the 'infusing teapot', was designed. The pots came in various styles, in order, it was stated, to suit the wants of the different classes. The interior was divided into two partitions, as shown, by a perforated strainer. The loose leaves were placed in the compartment nearest the handle, and the teapot filled with water. The unusually fine holes in the strainer were large enough to allow infused water particles to flow through, but small enough to arrest the broken specks of tea and dust. Thereby in the compartment beside the spout was to be found a clear infusion of tea, free of impurities and ready to be served.

The Teaette (1892)

For those occasions when a whole pot of tea was excessive, a
new utensil was created that was known as the 'teaette', which
provided a single cup of tea quickly and purely. The utensil
resembled a large silver spoon with a hinged lid, both bowl and
lid being perforated. The tealeaves were placed inside the bowl
and the lid was secured by means of a sliding ring along the
handle. A fresh cup of tea was instantly made by pouring the
boiling water through the teaette into a waiting cup.

The Carpet-Sweeping Machine (1887)

Many of the middle- and upper-class families across Britain
employed domestic servants to carry out housework such as
serving the tea and clearing away the crockery. In the days before
the motorised vacuum cleaner had been invented, sweeping
up cake and biscuit crumbs from the carpets was a tiresome
chore, and smaller rugs were often hung outside and beaten

clean. That was until the 'automatic carpet-sweeping machine' was invented in the United States and imported to Britain in the late 1880s. This new contraption consisted of a light frame running on four rubber-tyre wheels. Between two wheels at either end was a revolving brush which gathered up the dust and dirt as the machine was pushed over the carpet by means of the long driving handle. The sweepings were collected into two dustpans fitted in the frame of the machine. Upon pressing the triggers, the pans were automatically emptied without need for tilting or removing any portion of the machine, thus making it simple to use. For added convenience, the rubber furniture guard surrounding the wheels provided satisfactory protection against accidental damage to chair and table legs, which a careless servant might otherwise have caused.

The Foot Dustpan (1884)

Such domestic devices were not readily available to every family, and for those who could not afford the luxury of a carpet-sweeping machine – let alone a servant to operate it – a novel new dustpan was invented.

As every maid and housewife knew, the task of getting down on one's hands and knees to sweep an entire room with brush in one hand and pan in the other was by no means light work, but with the aid of one simple and effective device this arduous system was rendered practically effortless. The illustrated plan enabled the sweeper to employ a long broom while holding the pan by her foot. No stooping was required, and the upright position permitted of freer use of the broom. Owing to the arrangement of the patented appliance, the pressure of the foot caused the dustpan to lie flush with the floor, so that the dust always passed straight on to the pan. Furthermore, it could also be easily moved by foot to any part of the floor.

The Combination Spray Washer (1893)

More affluent Victorian families would have been able to afford
the services of several domestic servants, each employed to carry

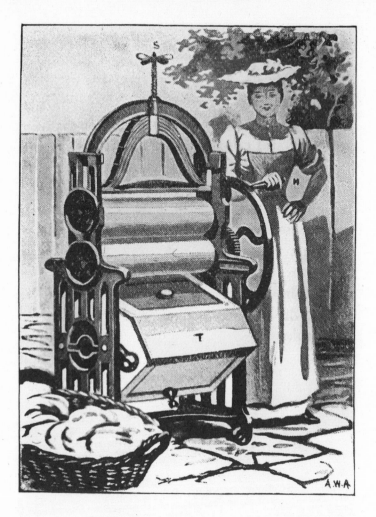

out their own specific duties. As the housemaid cleaned the house, the laundry maid busied herself with the washing and ironing of the household linen. One new appliance she would have found to be of great suitability was called the 'combination spray washer', which united, in a highly convenient shape, a clothes washer with a wringer and mangle, to wash, dry and then press the laundry all at once. The tub, T, in which the clothes were washed, was pivoted in such a way as to rock like a cradle when the handle, H, was turned, and very little force was required to keep it going. There were a number of wooden pegs in the bottom of the tub, which allowed the water to circulate, and the spray was dashed over the clothes from above. The fabrics were not strained or twisted, and even delicate muslins were not damaged in the process. In fact, the washer was designed to imitate the gentle, time-honoured method of washing clothes by hand in a stream.

Above the tub were two rollers through which the clothes were wrung after being removed from the wash, and the pressure of each could be adjusted by the screw, S, at the top. When the rollers were employed as a mangle to press the laundry flat, two detachable boards fitted into the frame above the tub.

The Gas Flat Iron (1884)

Prior to the introduction of gas in the early nineteenth century, families relied upon a solid cast-iron implement known as the flat iron to remove stubborn creases from the household laundry. These early irons were normally sold in pairs, so that while one was in use the other could be heated over an open

fire, ready for when the first had cooled. Though a thick cloth was frequently used to lift the hot iron from the fire, there were numerous reports of nasty burns to the hands and fingers. With this in mind, an American inventor introduced a safer flat iron that was kept hot by a gas flame, which burnt inside the hollow interior, away from the handle. In the figure on the previous page, P shows the flexible India-rubber pipe that supplied the gas to the burner inside the body of the iron. It yielded a uniform temperature, hence the linen was not likely to be singed, and it could be used for any length of time without cooling.

The Ironing Machine (1887)

To further assist with these labours, a new all-in-one 'ironing machine' was devised, combining the ironing table, as shown,

with a built-in iron that was situated above it. The appliance, which could be operated by hand or foot, was designed to regulate the pressure on the cloth, and gas was utilised in keeping the iron hot.

The Household Blowtorch (1887)

With the daily household chores made significantly easier by the introduction of these latest gadgets, inventors turned their attention to the more infrequent domiciliary duties. A convenient 'household blowtorch', for instance, was available to Victorian families by 1887, and was designed for thawing pipes, soldering, and burning off paint or wallpaper from the walls of the home. The handheld instrument was fuelled by a type of petroleum

spirit called benzoline, and a small pump enabled the user to supply a cold blast of air to the fuel in order to produce the blaze shown in the woodcut. The reservoir held 1½ pints of benzoline, capable of feeding the flame for three hours at full blast. The torch was also said to keep alight in strong winds when outdoor use was required.

The Combination Brush (1885)

A rather safer outdoor tool was the combined broom and hose, unveiled in 1885 for thoroughly cleansing pavements and doorsteps with minimal effort, rendering the usual pails and scrubbing brushes redundant. Patented in the United States, the traditional wooden broomstick handle was replaced by a metal tube, thus forming the key component of the invention. One end of this handle was connected to the main water supply via a flexible rubber pipe, while the other end terminated in a jet of water over the bristles of the broom, and thus the labour of first flooding the pavement with soapy water was done away with.

The Automatic Well Cleaner (1892)

In 1892, at Warwickshire's annual Royal Agricultural Society Show, a machine for automatically cleaning out wells was exhibited by an inventor from the market town of Wantage, Oxfordshire.

In an age when outbreaks of cholera and typhoid were all too common, and when the nation's rivers were constantly polluted by sewage, a fresh water supply was absolutely essential. Before the days of domestic tap water, Victorians relied heavily on rainwater for their everyday needs. While the lower classes were obliged to travel to the nearest communal well, some families were lucky enough to own a private well. This was indeed a useful commodity, but without careful attention a well was inclined to gather old leaves and sludge at the bottom, thereby contaminating the collected water.

The new well-cleaning implement, illustrated on the previous page, consisted of two pivoted tongs, A, with teeth at their extremities to catch the rubbish or stones, S, on the bottom. For the cleaning of mud and sludge, a pair of detachable scoops was employed in place of the prongs. The machine was lowered and raised by a rope or chain, E, attached to an eye, D. The jointed strut, F, allowed the prongs or scoops to close around the objects, so that they could be hauled up to the surface and disposed of.

This well cleaner from Wantage was also found to be extremely handy when personal possessions were accidentally dropped down the well.

The House-Warming Apparatus (1885)

A piped supply of fresh water was not the only commodity lacking in Victorian dwellings. Gas central heating may be taken for granted in today's comfortable households, but in Victorian times it was common knowledge that the hall was always the coldest part of the home. Chill winds frequently found their way into the unheated hall, its position generally being closest

to the outermost door. To solve this draughty problem, it was pointed out by a ventilating engineer that all that was required to send a genial warmth throughout the whole of a house was to thoroughly warm the lobby. This, he explained, could be achieved in a simple yet effective way with the use of the newly devised 'house-warming apparatus'.

A large coil of pipes, forming a square or oblong box, was placed against the wall of the hall, and connected by a flow and return pipe with a boiler in the basement or other convenient spot within the house. The boiler simply stood on the floor beside the chimney, like a stove, and was filled with water from a supply cistern that was placed at a higher level than the coil-box in the hall. The coil-table possessed a great deal of heating surface in a small space, so that a small amount of heat would amply warm the hall; the heated air gradually permeating the whole house and robbing the cold English spring of many of its terrors. Furthermore, placing the coil-box inside a painted metal case and fixing a marble slab on top could turn it into a handsome table.

The Fanned Rocking Chair (1893)

During the hot summer months those individuals who were not particularly fond of the heat craved an effective way of keeping cool. The image overleaf illustrates an ingenious American device by which the occupant of a rocking chair, in swaying backwards and forwards, worked a continuous rotary fan above their head. This was a purely mechanical arrangement of levers and pinions, which, by means of a screw, rotated a vertical shaft carrying the fan at its upper end. In sultry weather this device, which could be supplemented by a sunshade, proved exceedingly refreshing. The chair was also light enough to be carried inside.

The Combined Couch and Settee (1892)

An Army official named Lieutenant Colonel Davy invented a handy convertible settee for use in Victorian parlours and sitting rooms, particularly where space was at a premium. This novel piece of furniture conveniently combined a sofa, two chairs, and a small table or cup stand. The drawing opposite top represents it with the chairs and tables folded up, forming the couch alone. The scene below shows the furniture with the chairs opened

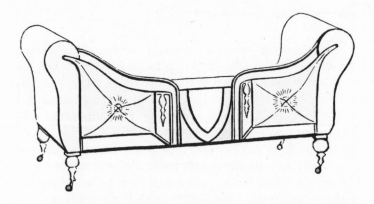

out, and the table spread between them. The combination was manufactured in different styles to suit a variety of tastes, and could, when fully extended, comfortably seat up to five or six people.

The Automatic Postman (1892)

A popular piece of traditional furniture was the writing bureau, at which many correspondents spent long hours. Before the invention of the telephone, the sending and receiving of letters was the usual method of long-distance communication, and people took much time and effort over corresponding with friends, business associates, and loved ones.

The postal service, however, was not a Victorian initiative. Messengers who delivered communications from one person to another across vast distances were employed in ancient civilisations, and in England it was King Henry VIII who established the Royal Mail back in 1516.

By the 1890s this timeworn institution was still being tweaked and improved, as were equivalent mail services across Europe. In Geneva, for instance, an apparatus was devised for effortlessly delivering letters and parcels to the upper storeys of high buildings. A communal letterbox was situated on the ground floor in the usual manner, but inside were compartments for each storey. Into these the postman put the letters and parcels, depending on which floor of the building they were to be delivered. A circuit was closed, and a current of electricity from a battery actuated a supply of water from a cistern at the top of the house. The water flowed into a cylinder acting as a counterweight to the letterbox, which began to ascend to the top of the building. On its way up various mechanisms were triggered which caused the contraption to automatically release its cargo at the appropriate storeys. When the highest floor was reached, the water escaped from the cylinder, and the letterbox returned to the ground floor to await the next delivery.

The Parcel Post Balance (1884)

For the use of persons wishing to send a large parcel by post, a spring balance was invented, which was graduated to weigh parcels up to the postal limit of 7 lb, and a scale displayed the corresponding cost of postage. A tape measure was also attached

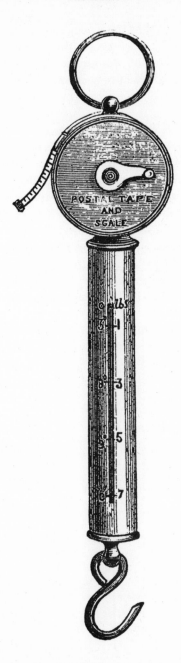

to the invention, enabling the sender to measure the dimensions of the parcel. Full instructions for the sending of these larger items were embossed on the measuring tape, and the whole contraption was made to fit conveniently inside one's pocket.

The Letterbox Annunciator (1887)

To further improve the delivery of letters, a gentleman named Mr Arthur Nahood devised a simple but ingenious plan for substituting the startling and disagreeable postman's knock for the more comforting sound of an electric bell. Its operation is shown in the following two diagrams, which represent a cross-section through the letterbox and door, and a back view of the letterbox with the electric circuit.

In the image below, A denotes the internal box into which letters were dropped, with its metal flap, B, fixed to the outside

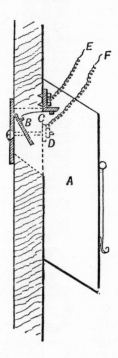

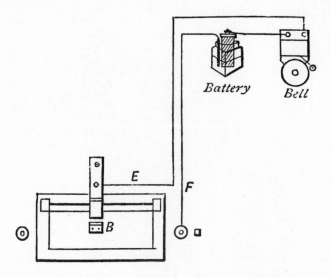

of the door. The postman pushed the flap open during the act of dropping a letter into the box. In so doing a metal contact on the back of B would touch a steel spring, C, that was in connection with a wire, E, which led to one pole of the battery through the electric bell. The flap, B, was permanently connected to the other pole by the wire, F, through the bolt, D, which fastened the letterbox plate to the door. When the flap, B, touched the spring, C, the electric circuit was completed, and the bell would ring to calmly announce 'Letters!' to the householders. When the flap fell back to its original position, the circuit was broken, and the bell ceased.

The Milk Receiver (1892)

In the days before refrigerators, householders would rely on one or two daily visits from their local milk vendor who transported his pails of fresh milk through the smoggy towns on a horse-drawn cart, or even slung across his shoulders. However, by the end of the century it was seen by some residents as somewhat inconvenient to have to repeatedly answer the door and allow the milkman to ladle his produce into a proffered jug. Therefore a

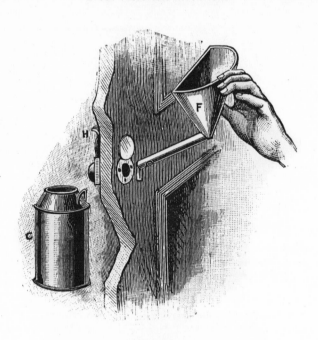

new plan of delivery was proposed whereby the milk was poured directly through a hole made especially for this purpose in the front door of the house.

A pivoted hook, H, was fitted to the inside of the door and arranged to open the hole when a milk can, C, was hung upon it. The milkman displaced the lid, L, and inserted his funnel, F, through the orifice, as shown. When the can was removed from the hook, the hole was automatically closed again, and a fresh and opportune supply of milk was available every morning for the residents to enjoy.

The Tell-Tale Milk Jug (1893)

As an added convenience for regular consumers, the new glass jug shown in the engraving was marked like a medicine glass

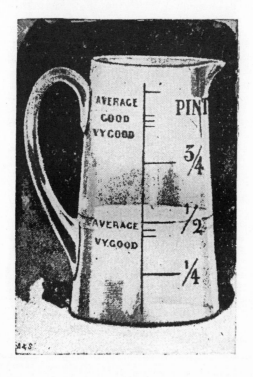

to indicate at a glance the quantity of milk received from the milkman. By means of three graduations, it also determined the quality of the milk – 'very good', 'good', or 'average' – by the thickness of the cream. The jug was thereby a means of selecting a good dairy for one's future purchases.

The Contractible Perambulator (1885)

Throughout history milk has formed the staple diet of young children, and during Victorian times, when infant mortality

was high, a healthy diet and proper care was key to the survival of youngsters. The majority of nineteenth-century doctors agreed that children should spend their days in a nursery room in the upper storey of a house, where ventilation was said to be better than on the lower floors. A supply of clean air was essential as dangerous airborne diseases hung about in the atmosphere, particularly in the smoky slums and overcrowded inner cities.

The perambulator was a real boon during the nineteenth century, for it enabled the nurse to take the children out for some fresh air without having to carry them. The great objection, however, to this contraption was that it was a rather cumbersome and inconvenient apparatus, so a new variant was invented by 1885, one which contracted, and was thereby capable of being driven down narrow passages or stored away in a small cupboard when not in use. This new perambulator was able to carry two children in its ordinary position, but by turning a key from the rear, the front section and wheels could be drawn underneath the back section, and the handle raised upright, as shown, thus enabling the whole vehicle to pass through very tight spaces.

The Brake for Perambulators (1885)

Accidents caused by 'prams' escaping from their attendants when travelling downhill, or due to their being unintentionally set in motion when left unattended on the pavement outside a shop, were of regular occurrence. To obviate these potentially lethal mishaps, the Halifax-born furniture manufacturer, Sir Thomas Henry Brooke-Hitching (c. 1858–1926), devised and patented a brake for perambulators, as illustrated. This useful little contraption was exhibited at the International Inventions Exhibition, a dazzling fair that was held at South Kensington in 1885, where the world's latest and greatest innovations were on display for the world to witness. Queen Victoria was patron, and the fair attracted nearly 4 million visitors during the six months it remained open.

To one end of the carriage's driving handle was attached a

lever, which hung freely when not in use, in such a way as not to inconvenience the nurse. This lever was connected by means of a chain to a pin that was fixed to an oscillating rod, whose bearings were secured to the body of the carriage. At each end of this rod were spoon brakes, which acted on the back wheels of the perambulator. When the lever was lifted, which could be easily done without taking the hand from the driving handle, the chain drew the pin upwards, thus rotating the rod to which the brakes were attached, and pressed them upon the wheels. A hanging ring was appended to the driving handle above the free end of the lever, and by slipping this ring over the end of the lever, the latter could be secured and the brakes thus held firmly over the wheels, so that baby and perambulator could be left unattended in the middle of the street without worry.

The Pocket Rack for Coats and Hats (1884)

When out and about, a new travel-sized contraption served as an opportune rack for a gentleman's hat and coat, where no other provisions were available. This useful device could be carried around with ease upon his person, ready to be brought out and set up at any location. As the engraving demonstrates, one hook carried the coat, and a spring clamp held the hat, while a second

hook suspended the whole. This handy contrivance was said to be particularly serviceable to cricketers and other sportsmen who were compelled to travel to the sports ground in ordinary attire.

The Umbrella Clock (1893)

In rainy Britain the umbrella was an object one was required to carry around at all times, and a creative inventor developed a basic improvement for the handle of this indispensable weather shield. His amendment came in the form of a neat little timepiece, as shown in the picture, at which convenient glances could be taken while hurrying along the puddle-strewn cobbles. The dial was only half an inch across, while the clockwork was

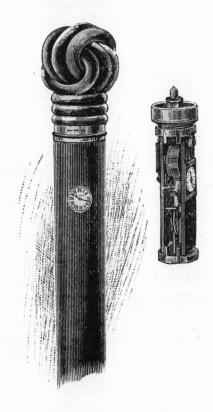

comparatively large, offering durability as well as accuracy. On those rare days when no rain was expected, the same clock could be removed from the umbrella and fitted into the head of a walking stick.

Time by Nightlight (1892 and 1884)

When the business of the day was complete, the Victorians would gather their families around the hearth to tell stories or play games, enjoy an evening meal and prepare to retire for the night. Electricity was a new commodity for nineteenth-century families, and many homes still relied on candlelight by which to see during the hours of darkness. For reasons of both economy and safety, these lights were extinguished before going to sleep, thus posing a great bother for those prone to waking during the long, lonely night, wishing to know the time.

The figure below illustrates a new type of candlestick that served as a night timekeeper, as the candle produced a nightlight for up to nine hours. The dial of its accompanying clock was

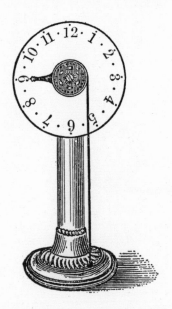

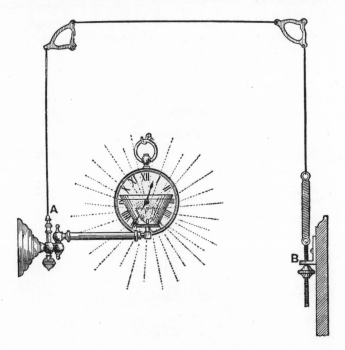

illuminated from behind in such a way so as to be visible from any position in a person's bedroom. The clock began to work as soon as the candle was lit. As it steadily burned it actuated a simple mechanism which imparted movement to the hand on the dial.

Similarly, an inventor named Monsieur Joyeux discovered a way of turning an ordinary watch into an illuminated clock, for displaying the time by night. As shown in the image opposite top, the device projected an image of the clock face on to a semitransparent screen. The source of light came from an ordinary oil lamp, which was placed in a small box fitted with an objective lens that could slide in and out a little, in telescopic fashion, to adjust the focus. The power of the lens and the distance of the screen determined the size of the reflected image of the watch, which was hung within the box behind the lens and illuminated by the lamplight.

The diagram below shows an alternative Victorian nightlight, this one powered by gas, which had the advantage of being controlled by a bell-pull arrangement (A, B), so that a person in bed could turn up the jet or extinguish the light in an instant. If a pocket watch was hung on the wall beside the nightlight, the flame would illuminate the time.

The Chanticleer Preventer (1887)

After a restful night, the last thing one wanted was to be woken early in the morning by the grating call of the cockerel. As many families kept these birds in their gardens, farmyards or orchards, the following 'simple' device was conceived by a cunning birdman in order to silence the irritating crowing of his cocks. The inventive gentleman loosely suspended a small wooden lathe about 18 inches above the unsuspecting bird's perch, and arranged it so that when the cock stretched its neck up to crow, it automatically actuated the device, and the swinging lath came gently into contact with its comb. This effectually stunned the poor creature into silence, according to the inventor, who stated that he owned a dozen birds, none of which ever presumed to crow until he let them out at a respectable hour. The regular use of this invention permitted the fellow to wake refreshed and alert, ready to begin a new day.

2

LEISURE

The Screw Lift for the Eiffel Tower (1887)

In 1887 Europe was abuzz with a sense of anticipation, for it had been announced that a world fair would be opening in Paris in two years' time to mark the 100th anniversary of the start of the French Revolution. L'Exposition Universelle would be a spectacular five-month-long event which had been years in the making.

The newly constructed entrance arch to this illustrious exhibition was hailed at the time as the tallest structure in the world. It took the form of a great iron tower and was named after its architect, Monsieur Gustave Eiffel (1832–1923). Completed in 1889, the sheer height of the Eiffel Tower necessitated the invention of a new type of lift, by which visitors could safely ascend to the top in one smooth and safe journey. This elevator was a kind of spiral railway, based on the simple principle of the screw and nut. Below the cage of the lift was placed a truck, or trolley, with three or more wheels running on as many rails, which ascended spirally like the threads of a screw. An electric motor caused the trolley to revolve, thus raising the cage above it. The cage itself was kept from revolving by fixed guides. Passengers inside the cage would not feel the spiral motion of the trolley beneath their feet, and all necessary precautions were taken to guard against accidents or too rapid a descent.

Unfortunately, despite the organisers' best efforts, the lifts were not completed in time for the grand opening, and visitors to the exhibition were obliged to walk all the way up to the very top of the tower in order to enjoy the views. This did not dampen the public spirits though, and by the end of the fair nearly 2 million tourists had made the ascent.

The Travelling Platform (1887)

As l'Exposition Universelle covered an area of Paris nearly 1 km square in size, one of the proposed strategies was to install a travelling platform, which would convey visitors throughout the different parts of the festival. It was reported that 3 km of rails were laid on the ground-floor levels, upon which trucks, powered by electricity, were run. These were low enough and sufficiently slow that passengers could alight at will and take in whichever area of the exhibition they pleased. Elevated cafés were provided on the railway platforms, and telephones enabled communication with the various attendants who were situated all around the site.

Unlike the tower lift, the railway was successfully completed, and was used by over 6 million visitors.

The Ferris Wheel (1893)

The Eiffel Tower was to l'Exposition Universelle what the great 'Ferris Wheel' was to Chicago's Columbian World Fair of 1893, that is to say, the chief wonder of the festival. This six-month-long event was a spectacular occasion, with the organisers pulling out all the stops to celebrate the 400th anniversary of Christopher Columbus's discovery of the New World. The astonishing Ferris wheel, the likes of which had never been seen before, was emblematic of the progress of science in the latter half of the nineteenth century. It was designed by the iron bridge engineer Mr George Washington Gale Ferris (1859–1896) of Pittsburgh, and is represented in the illustration, from which some idea of its size may be gathered by comparing it with the other buildings. The wheel measured 250 feet in diameter by 30 feet in depth, and revolved between two skeletal towers. It was

made of iron and steel, and the bearings were of aluminium bronze, each carrying a load of 600 tons. The axle was, at the time, one of the largest steel forgings ever made, being 32 inches in diameter, 45 feet in length, and weighing 56 tons. Cogwheels drove this enormous axle, gearing into endless driving chains that were actuated by a steam engine.

Though the wheel's inaugural rotation was operated without any carriages attached, this did not deter the workmen, who defiantly climbed the structure and took their seats among the iron spokes. The rim of the wheel eventually carried thirty-six coaches, suspended so as to always remain vertical as the wheel revolved. Rotating chairs were supplied within these cars, which were, in total, able to hold over 2,100 passengers at any one time.

This original Ferris wheel, upon which every subsequent variation was based, attracted a great deal of attention from the thrill-seeking visitors to the fair. The ride took ten minutes to complete one revolution, and offered passengers the most breathtaking views of Chicago once their car had reached the top.

Mr Ferris had been determined to 'out-Eiffel' Mr Eiffel, but unlike the famous tower, which even Herr Hitler couldn't take down, the iconic wheel was not permitted to remain standing. At the close of the exposition the wheel was transported to St Louis for the grand opening of the Louisiana Purchase Exposition in 1904, which marked the centenary of the United States' acquirement of this region from French control. The ride was ultimately demolished in 1906, and its memory confined to the pages of history.

The Gravimotor (1884)

In 1884 a very different type of pleasure wheel was invented, which was termed the 'gravimotor'. This self-impelling roller-skate was invented by two Canadians named Messrs Tomas P. and James B. Hall. The boots were strapped to the feet just like a normal pair of skates, but as soon as the weight of the body pressed down on the footrests, the wheels began to revolve on their own, propelling the rider forwards. No action was required on the part of the rider, for the device was entirely automatic, allowing the user to relax and enjoy the scenery as they went on their way.

The motors beneath the footrests were each composed of two wheels, revolved by a spring in tension, by the intervention of toothed gearing. As there was a motor attached to each foot, it was only necessary for the rider to put their weight on to one skate at a time. On relieving the footrest from the weight of the body, the wheels spontaneously ceased to revolve, and the passenger came to a halt.

The Rowing Machine (1885)

For many years a type of apparatus had existed which enabled people to enjoy the exercise of rowing with need for neither boat nor water; but such machines were generally objected to on account of their being stationary. In an attempt to resolve this predicament and offer the public a more leisurely form of exercise, an American inventor devised a new kind of rowing machine. His apparatus consisted of a boat-shaped car on wheels, complete with oars and a continuous, circular track, which could be easily laid at any location. The physical act of rowing impelled the vessel along the track with a certain degree of exhilarating impetus, which had been impossible to obtain using fixed machines. Thus the benefits of exercise could be reaped while the rower watched the world go by.

The Gymnastic Treadmill (1893)

Another type of exercise machine, as shown in the illustration, was brought out in France and was designed for the purpose

of affording a gentle workout akin to walking up a slope. This 'gymnastic treadmill' consisted of a series of rollers, R, covered with soft carpet, and inclined at an angle to the horizontal. Two upright supports, H, for grasping with the hands to steady the operator, rose like the back of a chair. In treading on the rollers these carried the feet downwards, and in order to keep one's position it was necessary to walk rapidly, or even run, as the rollers revolved. It was possible to graduate the extent of movement by raising or lowering the frame of the rollers by means of the screw, S. The higher the inclined plane was set, the more vigorous was the exercise.

The Candle Motor (1893)

Children were often kept entertained with simple toys and games such as dolls, dominoes, or the traditional 'cup and ball'. In France, a simple but imaginative – and withal instructive – little toy, known as the 'candle motor', was invented, which any *père* could make for the amusement of his *enfants*. To create the

toy, the ends of two pins were heated and then inserted at right angles into opposite sides of the middle of an ordinary candlestick. These pins formed the axis of the motor, and rested on the edges of two wineglasses, thus supporting the weight of the candle between them, as shown.

To simulate the motion of a seesaw both ends of the candle were lit, and plates put underneath to catch the drippings. A drop of wax would fall from one end and upset the equilibrium of the candle, causing it to rock. Very soon an oscillatory motion was set up, which only ceased when the candle was extinguished. Cardboard figures attached to the ends of the seesaw with thick wire provided the finishing touches.

The Prison Puzzle (1887)

While the French were adept at producing quirky little toys and games, back in Britain the renowned astronomy writer

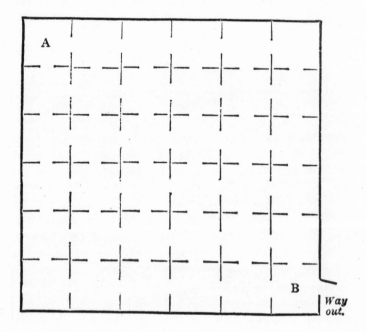

Mr Richard Anthony Proctor (1837–1888) devised a rather challenging brainteaser. He believed it had only one solution, but never elaborated.

The figure represents the plan of a prison containing cells which communicated with each other, as shown. A prisoner in cell A was offered his freedom if he could make his way to B by passing once, and only once, through each of the thirty-six cells.

How was he to do it?

The Underwater Spyglass (1887)

British prisoners weren't the only ones deprived of their freedom in the late nineteenth century. Though the slave trade had been abolished in Britain (an Act having been passed by Parliament in 1833), captives were still being bought and sold in other parts of the world, particularly in Africa and the Arab kingdom.

In an effort to earn a healthy profit, some masters taught their slaves how to dive in order that they might explore the Red Sea in search of valuable mother-of-pearl. To better spot the pearl shells on the seabed, and thereby collect more treasures than their competitors, the slaves were instructed to make use of a new spyglass. This invention consisted of an empty tin with a pane of glass fitted to its bottom. The box thus made was partly submerged, glass bottom downward, so the diver was able to gain a clear view through the glass to the sand below, unobstructed by surface ripples.

Once they had collected their bounty the slaves took two thirds of the profits back to their masters, leaving a third to the owner of the boat that had taken them out to sea. The shells, after being cleaned, were packed in barrels for export. Some went to Trieste in Italy, some to Havre in France, and some to London. Others went to Bethlehem, where they were carved and sold to pilgrims for a princely sum.

To clearly observe the beautiful coral reef sea gardens in the Bahamas and other West Indian islands, travellers and sightseers began purchasing the same sort of underwater spyglasses used by the slave divers, and what dazing sights they must have witnessed.

The Water Bell (1885)

A great many people during the nineteenth century appreciated the beauty of crystal-clear waters, and until 1885 Victorian water bells – popular outdoor features in middle- and upper-class gardens – had hitherto been made by allowing a supply of water to flow from a small circular orifice, or by shooting the liquid jet against a disc of polished metal with a slightly elliptical rim. The bells thus produced were generally deemed to be lacking in transparency, and a new type of bell, formed by two opposing jets of water, was considered to be a great improvement. This new design was said to be the invention of a Frenchman named Monsieur Bourdon.

To produce the bell, the basin was filled to the level of overflow with water, the stopcock was gently opened, and a ball of water was formed between the tubes, which gradually enlarged into the bell-shaped fountain of the engraving. By reducing the aperture of the stopcock a little, the shape changed into that illustrated.

It was possible to slit the bell by pointing a thin copper wire into its top, and by the hole thus made, a statuette, a lighted candle, a birdcage, or other article could be introduced to the display without fear of it getting wet.

The tubes used by Monsieur Bourdon were 20 mm in diameter, but by employing larger apparatus, water bells of several yards in diameter were transformed into crystal tents, under which people were able to walk. In hot summer weather such alcoves were thought quite refreshing, and formed the perfect setting for garden parties.

The Sandwich Indicator (1892)

Sandwiches were becoming familiar offerings at alfresco gatherings. By 1892 the content of a typical Victorian sandwich was no longer limited to ham, beef or tongue. In fact, a wide variety of delicious new fillings, such as egg, cheese, and minced goose, were now entirely fashionable. Even so, traditional party-goers were not accustomed to such a diverse range, and some form of indicator was required so that they might know what meaty mysteries lay between the slices of bread. To their relief, a lady named Mrs de Salis envisaged a handy utensil, which consisted of a small, prettily stamped metal dagger that was to be thrust into a plate of sandwiches. At its top it supported, by way of a tiny chain, a little frame, into which could be slipped a card indicating the flavour of sandwich to be found on the dish, hence eliminating any unfavourable surprises.

Soil Heating (1884)

As well as picnicking, another popular outdoor hobby was gardening, and many families, especially those who lived in rural areas, chose to grow their own food. By the 1880s it was expected

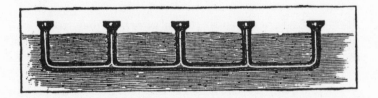

in certain horticultural circles, particularly on the Continent, that an innovative new plan to force the growth of vegetation by directly heating the soil, rather than the plants, by means of hot underground pipes, would enter widespread use before the turn of the century. The woodcut represents the pipes running beneath the soil, which at intervals rose to the surface to heat the air. Steam was generally employed to warm the pipes, but it was also reported that an Italian businessman named Sig. Francesco Cirio (1836–1900) successfully utilised the hot springs of Acqui by circulating their waters in earthenware pipes laid underneath the ground. The water was pumped from the town baths to Sig. Cirio's nearby gardens, where 10,000 asparagus, 4,000 chicory, and 4,000 lettuce plants were said to have been effectively forced in this manner. This yield was extremely beneficial for Cirio, who had, as a young man, achieved distinction by inventing tinned vegetables. He subsequently founded a successful food-exporting business, and discovered he could transport his canned products a greater distance without the contents being spoiled.

It wasn't long before the Japanese learned of this effective technique for forcing vegetables, and they copied Cirio's method by utilising the springs of Tokyo in a similar manner. It was also proposed to harvest the hot gases found in volcanic districts for the same purpose.

The Garden Waterer and Roller (1885)

In 1885 a handy new gardening tool was invented that combined the ordinary garden roller with the watering can. The engine of the new device was mounted upon two rollers, and a pump was attached to the tank for forcing the water into a jet. While this proved very useful for spraying sturdier plants, such a powerful stream of water was prone to damage carefully arranged or delicate flowers, or even wash the soil away from their roots. This difficulty was overcome by a siphon arrangement, the workings of which can be seen in the illustrations.

The opposite image shows the engine as arranged for watering shrubs and hardy plants. The tap, C, was closed, and as the pump

was worked, the water was forced through the pipe, A, to the jet, B, by which it was diffused. The rollers upon which the engine was mounted enabled the machine to be moved across lawns, along paths, and even round corners with ease.

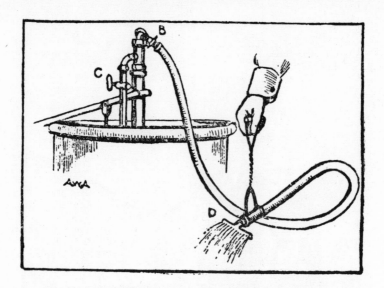

For the watering of delicate flower bedding, the jet, B, was removed and a hose attached to the nozzle as shown above. The tap, C, was then opened, and the water flowed gently through it to the hose until the tank was empty. A suitable sprinkler, D, was supplemented at the free end of the hose.

The Gardening Tool (1885)

When tying up plants at a height which could not be easily reached by the gardener, there was often considerable difficulty in holding the cord. To alleviate this problem, an American inventor patented a new tool, as displayed in the opposite image. The centrepiece was T-shaped, and its upright could be adjusted to any length. Slight spring clips were attached to the points of the crosshead and the two projecting rods, all opening inwards. The cord, having been tied in a running noose, was secured by the clips and passed round the plant at the desired point by means of the tool, and, on being pulled, escaped from the clips and drew the parts of the plant together.

The Self-Cleaning Garden Rake (1885)

Victorian horticulturists who experienced the bother of having to remove leaves and cut grass from the teeth of their garden rakes found the following addition to their outfit wholly advantageous. A self-cleaning garden rake was provided with a frame of bent wire attached to the head of the rake by means of a spring, which kept the frame away from the teeth while the rake was in use.

When an accumulation of leaves and dirt was to be removed, the gardener was simply required to press his foot on to the frame, which was pressed down over the teeth, clearing them at once from debris.

The Ammoniaphone (1884)

Though beautiful English country gardens were the envy of horticulturalists the world over, the infamous British weather was not so coveted. Outdoor pursuits were often deemed more agreeable in warmer European countries than in rainy Britain. Italy, for example, boasted breathtaking scenery for sightseers to admire, and its favourable climate was said, by one Scottish chemist, to explain why the Italians were so celebrated for their exquisite singing voices. Over the years various theories were proposed as to why these countrymen had been gifted with such sweet, melodic tones, and in 1884 Dr R. Carter Moffat felt he was at last able to solve this conundrum. After spending a considerable amount of time studying the Italian climate, he discovered that Italy's air and atmospheric dew possessed special qualities, specifically that their levels of free ammonia and peroxide of hydrogen were far greater than anywhere else in the world.

Upon his return to Scotland he set about constructing a metallic apparatus called an 'ammoniaphone', which was used as an inhaler to imitate the soft balmy atmosphere of the Italian peninsula. This thin, tubular instrument, which was about 25 inches long, contained absorbent material saturated with peppermint oil, peroxide of hydrogen, and condensed ammonia. A deep whiff of the contents, if taken regularly, was said to Italianise and greatly improve one's voice, producing a rich, powerful and melodious tone of extraordinary clearness and range.

According to an advert featured in *The Yorkshire Post and Leeds Intelligencer*, published on 11 September 1884, the instrument's use was recommended for 'vocalists, clergymen, public speakers, parliamentary men, readers, reciters, lecturers, leaders of psalmody, schoolmasters, amateurs, church choirs, barristers, and all persons' who were required to 'use their voices professionally'. Not only this, but the marvellous ammoniaphone had allegedly been found to be an invaluable remedy for most pulmonary diseases, including coughs, colds, bronchitis, asthma, consumption, loss of voice, affections of the throat and chest, sleeplessness, and even temporary deafness.

On many occasions, singers had publicly demonstrated the value of the inhaler for improving the range and quality of their voices, and Dr Moffat was said to have received over 5,000 glowing testimonials from satisfied customers. Even the Prince of Wales was an alleged admirer. In short, the *Sheffield Daily Telegraph* declared that the ammoniaphone was one of the most notable inventions of the day, which had all the promise of triggering a 'revolution in vocalisation'.

It was perhaps unbeknown to the clergymen, schoolmasters, and amateur singers who used this wondrous contraption that ammonia is a corrosive gas, and exposure to this chemical in significant concentrations has been found to cause coughing fits as well as nose and throat irritation. Continual contact with the gas often results in burning sensations to the eyes, nose and throat, leading to blindness, lung damage, or even death. Similarly, peroxide of hydrogen is not recommended for human ingestion, not least because the vapours are said to be explosive.

The Aeolia Harp (1892)

Music was a popular form of entertainment in the days before radio and television were invented. Savouring the melodious

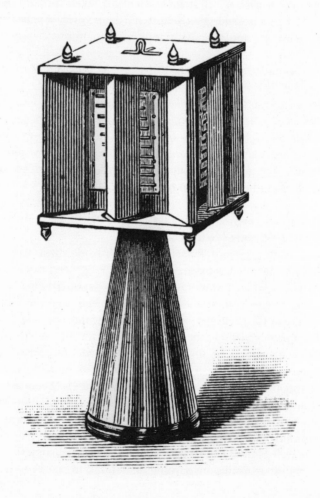

tunes of a well-played instrument was considered to be a most delightful way of spending a summer's afternoon.

The 'Aeolian harp' was a musical instrument invented during ancient times. It was named after Aeolus, the Greek god of the wind, owing to the fact that it was played by the gentle motion of the breeze. A traditional Aeolian harp measured about 4 feet high by 18 inches wide and 9 inches thick, containing between seven and twenty-one strings. These were apt to go out of tune fairly easily, so the Victorians conceived a new version of this classical instrument. In the 'Aeolia', as it was termed, there were no strings, but eighty sensitive metallic reeds divided into harmonic chords of twenty notes each. The apparatus is shown in the engraving, and stood 20 inches high by 8 inches wide and deep. As the upper extremity of the instrument revolved in the breeze, it played one chord after another, no matter from which direction the wind was blowing. All the reeds stayed in tune and sounded into one common organ tube, which harmonised the chords.

The instrument was designed to hang from a tree or be fixed on a post in a village green. For the greatest atmospheric effect, it could sometimes be found placed beside a ruined castle.

The Geological Piano (1885)

It was not just the melodic sound of the wind that caught the Victorians' ears. It had long been known that certain flint stones possessing special properties emitted a musical note when struck with other flints. Monsieur Honoré Baudre was a nineteenth-century French musician whose greatest pleasure in life was collecting these sonorous stones of prehistory. He lived in a village close to the River Indre in central France, and devoted more than three decades of his life travelling the world in search of perfectly pitched flints, which he painstakingly tested one by one, in order to construct his magnificent 'geological piano', deemed by some to be the strangest piano in existence. His flints were chosen according to their pitch, and were suspended in a horizontal row by their two ends, in a regular musical scale. An elegant metal framework supported the stones, which were hung by thin wires. A sounding board was placed a fraction of an inch below the row of stones, and tunes could be played by means of two smaller flints held in the hands, which rang out the notes in melodic majesty.

Though it was a remarkable piece of craftsmanship, it transpired that Baudre's geological piano was certainly not unique. The lithophone is a percussion instrument made from rocks and stones, similar to the xylophone, which has been in existence for centuries. Decades before the invention of Baudre's piano, Mr Joseph Richardson, a talented stonemason from Cumberland, began building a lithophone that became known as the 'Richardson Set' – a task that took him thirteen years to complete. His remarkable instrument was constructed from impeccably tuned sonorous flints that had been hand-picked from the enchanting 'Musical Stones of Skiddaw': a collection of rare rocks found in the Lake District. By the 1840s Mr Richardson and his three sons had set up a rock band known as 'Joseph Richardson & Sons and the Rock, Bell & Steel Band', and they set off on what was initially supposed to be a three-week tour around Northern English towns to perform their music. However, their unexpected success meant that they did not return home again for three years. In 1848 they were even invited to Buckingham Palace to perform for Queen Victoria and Prince Albert.

The Xylophone (1885)

The xylophone was played in ancient civilisations in one form or another, but it wasn't until about the year 1830 that a Russian Jewish musician named Michal Josef Gusikow (1806–1837) brought it to the attention of a Western audience. He invented his own version of the instrument when a lung condition rendered him too weak to continue playing the flute, and like the Richardsons, the Gusikow family went on the road, playing their novel instrument to an appreciative public. In the 1880s the xylophone finally came to the forefront in French orchestras, and was played in the manner shown opposite.

As the name implies, the xylophone was 'an instrument of wood and straw', and the wooden rods of varying length were strung together by cords, so as to form the trapezium figure above. The frame was laid on bands of straw to amplify the sounds, rendering them stronger and purer, and the notes were produced by striking the pieces of wood with a couple of small hammers.

The Music Writer (1887)

For the actual composition of pieces of music, the engraving overleaf depicts a little instrument that was worked on a rotary wheel principle. This new machine was invented for the purpose of writing musical characters, thus proving itself beneficial to professional and amateur musicians of the day. In order to use

it, the composer held the handle and rotated the type wheel until the desired musical character reached the bottom, just above the paper. As the wheel was spun, a roller automatically inked the characters. Then, by pressing down on the handle, the musical note was brought into contact with the paper.

The design had the great merit of simplicity, so that anyone could use it, and unlike its typewriting counterparts, its price brought it within the reach of almost everybody. Furthermore, the apparatus was so small that it could be carried around to lectures, offices, or music venues in an ordinary coat pocket.

The Folding Music Stand (1885)

By 1885 a firm from Newbury, Berkshire, had brought out an inspired new music stand that folded up into a portable form without having to be taken to pieces. The accompanying

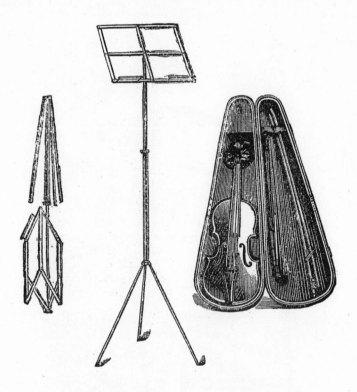

woodcuts illustrate the stand when erected, when partially folded up, and when inserted in the lid of a violin case for transport. All screws, bolts and springs, which were troublesome in ordinary stands, were dispensed with, as the stand folded up by the hinging of its parts. Its height could be varied at will to suit any performer, and as it was constructed of thin metal and tubing, it was very light and durable.

The Reading Chair (1892)

Reading was another popular hobby in the nineteenth century. With the evening's entertainment in full swing, stories could often be heard drifting down the corridors of a typical Victorian dwelling. The illustrated reading chair was an improved model

of one that had long been used by the Duke of Wellington (1769–1852) at Walmer Castle in Kent. The back – or rather, the front – was so formed that a person could sit astride the chair and comfortably lean their arms on the shelf with the book before them, in order to recite fireside tales to a captive audience, or to simply relax and read alone during quiet times.

The Whispering Machine (1885)

The concept of 'automatic talkers' was by no means novel during the late Victorian era. Indeed, since the invention of the photograph in the early 1800s this had been an oft-discussed notion, though the requisite technology was not yet available to turn this dream into a reality. However, in 1885 it was reported that the celebrated inventor Mr Thomas Edison (1847–1931) intended to try to develop 'electrotyped phonographic books' which would read aloud, thus saving the reader's eyes from needless strain.

The 'Whispering Machine', though never realised, was an exciting proposal that would have surely held the technophile daydreamers of the era in awe. These tiny machines were intended to be placed close beside a person's ear, such as inside their hat, in order for a pre-recorded voice to discreetly narrate stories to the wearer without disturbing their peers. With this concealed contraption a man might take a walk along a busy street while the book of the season was whispered into his ear, helping to pass the time in a pleasing manner, but it would take another 100 years for technology to catch up with him.

3

FASHION, ART AND DESIGN

Health Clothing (1885)

The late Victorian fashion industry was booming. Advances in sewing and lace-making machines during the middle of the nineteenth century had made the lives of costumiers considerably easier, and more elaborate clothing designs were being produced for a much cheaper price. Trade from the colonies was also prosperous, which meant that beautiful and exotic new material was being imported to Britain, ready to be transformed into the finest suits and most exquisite dresses imaginable. Communication links were also better than ever, and by the 1880s all eyes were on Paris, Europe's capital city of fashion.

While the well-dressed women of England admired the latest fur cape and the must-have tea jacket, over in Germany a professor of physiology began considering clothing from a scientific point of view, rather than an aesthetic one. In 1885 Dr Gustav Jäger (1832–1917) devised a new range of 'health clothing', which was met with much appreciation in his homeland. The doctor's theory was that unlike dead animal tissues, dead vegetable matter did not dispel the natural secretions of the skin, and this type of fabric was thus counter-beneficial to one's health. His discoveries alluded to the fact that a person was less likely to be predisposed to disease when their body contained a low amount of fat or water, and that their nervous activity, or, in other words, their sensitivity to touch, was greater. His experiments showed that, after taking a Turkish bath to reduce the moisture of the body, a gain of 13 per cent in nervous activity was registered by the subjects.

Holding as he did that linen and other vegetable-based clothing was not adept at dispersing moisture from the body, Dr Jäger began manufacturing a collection of health attire made entirely from materials derived from animals.

His design for a woollen dress consisted of tight-fitting stockingette underclothing made from pure undyed wool, fastened over the shoulder, and of double thickness over the breast. His jacket was also double-breasted, buttoned well up to the throat, and contained no lining or padding. The same design was applied to the trousers, and the vest was made in the form of flaps that were sewn to the interior of the coat.

Dr Jäger's experiments had shown that colds, rheumatism, lumbago, and other such complaints were caught by the sudden rush of cold air to a particular part of the body, so inside the jacket sleeves and trouser legs were sewn pieces of material that were tightly fastened round the limbs to prevent up-draughts. Instead of a traditional starched linen collar, Dr Jäger substituted one of unstarched white cashmere, which was not only comfortable, but a safeguard against sore throats.

The wearer's feet were clad in pure woollen socks with individual divisions for each toe. The upper part of the boot was made of felt, while the lower part consisted of porous leather, and the inner soles of perforated leather and layers of felt; thus the boot was quite porous, and the feet were kept as clean and pure as the hands.

By doubly protecting the entire body in this way, the wearer's blood vessels were thought to have been stimulated, and as an even temperature was maintained throughout, there was little need for an overcoat. It was also claimed that as well as being warm in winter, the clothes were cool in summer.

The same health principles were adopted for night clothing, whereby bed sheets were made of wool, camelhair, or white cashmere. The mattress and pillows were stuffed and covered with wool, and Dr Jäger insisted that it was safe for a person so protected to keep his window well open at night for ventilation purposes. However, because the Victorians knew all too well that open windows admitted potentially harmful draughts on to the exposed head, it was advised that some kind of screen should be employed to prevent this.

The International Health Exhibition of 1884 consisted of a series of talks, lectures and demonstrations on the subject of public health, including food, clothing, the dwelling-place, heating, lighting, and ventilation, and samples of Dr Jäger's revolutionary new clothing range were exhibited. Subsequent to the appreciative reception his designs received, a depot for the sale of the doctor's health clothing was opened in London under the name of 'Dr Jaeger's Sanitary Woollen System'. After several takeovers and rebrands, the company still operates in Britain and is known simply as 'Jaeger'.

Straw Shoes (1887)

In 1887 an American named Dr Macgowan suggested the introduction of straw shoes to the nursery for the use of children. According to him, footwear made from either rice-straw or mat-grass gave freedom to the feet, allowing fresh air to circulate, and better still they cost just a few pence per pair. Not only did they offer health benefits, the shoes also provided employment opportunities for the needy, as they were made by 'aged people, too feeble for more active employment'.

Asbestos Hat Linings (1885)

While the likes of Drs Jäger and Macgowan were busy inventing clothes they claimed were beneficial to one's health, hats with a thick lining of asbestos in the crown were being readily manufactured in the United States and sold to a public oblivious to the risks. As this wondrous, versatile material was a well-known insulator, its use for this purpose was thought to be an effective way of keeping the wearer's head warm.

The Industrial Revolution had seen asbestos factories popping up all over the place, with new uses for this economical substance consistently dreamed up by would-be entrepreneurs. Little did these eager speculators realise how incredibly toxic this versatile material was. The first death to be caused by close contact with asbestos wasn't recorded until after the turn of the twentieth century, and the first diagnosis of 'asbestosis', the fibrotic and inflammatory complaint of the lungs caused by the constant inhalation of asbestos fibres, was made as recently as 1924.

Adjustable Braces (1887)

A new version of a rather less dangerous item of Victorian clothing was introduced to Victorian England in 1887. The figures show back and side views of some newly patented gentlemen's braces. The outstanding feature of this accessory was the cross-tree arrangement, which, especially at the back, gave a very pleasing result. By wearing this device all dragging of the trousers at the heels, as well as accidental exposure of undergarments, was avoided. Furthermore the braces were adjustable, thus allowing the wearer to arrange them to his own comfort.

The Pencil Suspender (1892)

To complete the gentleman's outfit, the little device for handily suspending pencils, illustrated herewith, was designed to be worn as a locket either on the dress or on the chain of the pocket watch. This was particularly convenient for the businessman who often required the use of a writing implement at short notice, when one was not always to hand. The pencil was hung by a cord, which, after being drawn for use, flew back of its own accord like a spring tape, taking the pencil with it once its services were no longer required.

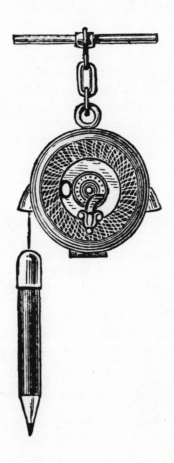

Electric Jewels (1884)

For the lady, Monsieur Gustave Trouvé (1839–1902), the famous Parisian electrical instrument maker, devised a number of very pretty and highly inventive personal ornaments, such as breast-pins, diadems, and brooches, which, while resembling the

traditional jewelled articles, possessed a still greater brilliance. Miniature electric incandescent lamps were concealed in the centre of the trinkets, emitting twinkling lights from behind the gemstones when turned on. Wires led from the bulbs to a small pocket battery, enclosed in an ebonite case, which was secreted in a convenient place upon the person. The power from the battery was said to last for up to thirty-five minutes.

The top and bottom left images on the previous page represent a brooch and a scarf pin. The bottom right image on the previous page depicts an unusual diadem topped with an electrical star light, in which the light was fed by two cells, C C, of the miniature battery, carried as part of a belt buckle worn round the waist, the current being carried to the lamp by fine, silk-covered wires. A small switch, S, beside the buckle allowed the wearer to turn the light on their head on or off at will.

The Illuminated Rose (1887)

Another type of sparkling accessory is represented in this diagram, which illustrates a small pocket apparatus for the

electric illumination of fresh flowers that adorned the hair or the dresses of the most dedicated followers of ladies' fashion. The invention, similar to a modern-day battery pack, consisted of an ebonite case containing three voltaic cells. It measured 4 inches high by 1 inch thick and 3 inches wide, and could be carried unseen in the pocket. Flexible silk-covered connecting wires led from the poles of the battery to the tiny electric lamp, L, which was placed in the centre of the flower, as shown. Like the electric diadem, a switch on the box enabled the light to be turned on or off, as the user desired.

The Buttonhole Flower Holder (1885)

While these illuminated roses looked positively charming when arranged in a lady's hair, flowers were more commonly pinned to the breast pocket of one's jacket. Victorians who regularly sported a buttonhole were no doubt interested to learn of the invention of a new kind of flower holder. It consisted of a case of silver, and was to be attached like a brooch to the coat or dress. The back of the case was hinged to the front, and the whole was kept fastened by a ball clasp fastening, common in some ladies' purses. On the inner side of the back of the case two tiny pins had been soldered in. To insert the flowers, the case was opened, and the pins served the purpose of keeping the stalks of the flowers in a suitable position once the case had been closed. This was the special feature of the holder, as it enabled the wearer to artistically arrange the buttonhole before pinning it on to the attire.

When the holder was not in use, it could be kept, just like a piece of jewellery, in an accompanying case that had a fine velvet lining sewn inside.

Curious Needles (1884)

Though the common sewing needle was an article of great practicality to women of the nineteenth century, its artistic qualities were of little significance. In July 1883, however, an exposition dubbed the International Exhibition of Needlework was held at London's great Crystal Palace, where two particularly eye-catching needles were displayed.

One had been made at Redditch and presented to Queen Victoria. It took the form of a sort of miniature Trajan's Column, engraved from head to tip with scenes from Her Majesty's life, so fine that a magnifying glass was required to see them. Moreover, the needle was hollow and could be opened up, whereupon it was found to contain a number of even finer needles, like Russian dolls, on which further scenes were engraved.

The other needle on display had been presented to the German emperor, William I, during a visit to a state-of-the-art needle factory at Kreuznach. On being shown some exceedingly delicate examples, His Imperial and Royal Majesty expressed his surprise that needles so fine could actually be threaded, prompting the foreman of the boring department to request a single silvery hair from the beard of the emperor. With trepidation the Kaiser complied, and was astonished when the skilled workman proceeded to bore a hole directly through the tip of the hair, before looping an even finer thread through it. The foreman returned this unique artefact to the emperor, who kept it as a prized souvenir for the remaining years of his life.

A Marvel in Clocks (1887)

It wasn't only miniature artefacts causing a public sensation in the nineteenth century. On 27 January 1887 it was reported in *The Star* newspaper that, after decades of toil, Herr Christian Martin, a notable German clockmaker from Villingen in the Black Forest, had finally succeeded in inventing a magnificent 'clock of clocks', the likes of which had never been seen before. This wondrous mechanism was said to surpass the elaborate timepiece of Strasburg Cathedral, and even put the celebrated Swiss astronomical clock of Bern Tower in the shade.

Martin's masterpiece was said to stand 12 feet high in a Gothic case. It was designed to display not only the seconds, minutes, hours, days, weeks and months, but also the seasons and years, including leap years, until the very last second of the year 99,999 AD. Moreover, the clock showed the time in different parts of the world, as well as the phases of celestial bodies such as the sun and moon, and its musical works were said to resemble the sweet and delicious tones of a flute.

If that wasn't enough, this amazing clock also depicted a series of moveable figures representing the story of the Creation and the Stations of the Cross, as well as various pagan mythologies. A miniature skeleton enacted Hans Holbein's *Dance of Death*, and at night the figure of a watchman came forward and sounded the hour on his horn. A cockerel crowed at sunrise, and a cuckoo appeared once a year on the first day of spring. At a certain hour of the day the tiny carved figure of a sacristan materialised in order to ring a bell in his church spire before kneeling down and folding his hands in prayer. Another section of the clock depicted the twelve Apostles, Shakespeare's Seven Ages of Man, and the twelve signs of the zodiac.

To top it off, the clock came with a free-of-charge 10,000-year guarantee, after which time, according to adverts for the clock, 'its mechanical works will have to be changed'.

The Crystal Chair (1884)

The English were just as adept at fashioning wondrous creations. In 1884 a manufacturing firm in Birmingham was commissioned to design a dazzling chair made entirely out of crystal, to serve as a throne for an Indian potentate. It was composed of crystal columns and polished facets, and was covered by a crystal canopy, which was illuminated from beneath by three electric incandescent lamps. The effect was said to be very fine indeed.

The Largest China Vase in the World (1884)

At around the same time, a well-known firm of Staffordshire potters, 'Messrs Brownfield and Sons', was hired by an unknown benefactor to produce a china vase that would not only become the largest the world had ever seen, but also a unique work of fine art. A special oven was constructed to fire this magnificent piece, and by 1884 the vase was complete. Its chief spectacle was a beautiful globe supported on a pedestal, which rose from a square plinth. Round the centre of the globe ran a frieze divided into four panels representing each of the four seasons, on which a number of cupids were seen occupied in seasonal pursuits. Similarly, the pedestal incorporated a splendid frieze, on which were represented as many as sixty cupids, all occupied in rural work. The globe was surmounted by a figure of Ceres, the Roman goddess of agriculture, who, aided by even more cherubim, was occupied in showering her gifts of fruitfulness and plenty upon the Earth.

The globe was a subdued celadon green, while the figures were made from china bisque, and the other decorations from white glazed china. Though it was the company's art director, Mr Louis H. John, who designed the vase, the figures were modelled by the distinguished French sculptor Monsieur Albert-Ernest Carrier-Belleuse (1824–1887), in a unique and world-record-breaking collaboration. From the pedestal to the top of the figure of Ceres, the vase stood 11 feet high, and the diameter, including the ornamental figures, was 6 feet 4 inches. Notwithstanding its massive proportions, this thoroughly artistic vase was regarded as elegant and chaste.

Before the vase left the pottery at Cobridge for the International Exhibition at the Crystal Palace, it was put on display for just one day for the local townsfolk to view, and as many as 25,000 persons availed themselves of the opportunity. Some time later, the organisers of l'Exposition Universelle invited the manufacturers to exhibit this wondrous creation at the Parisian world fair in 1889. Sadly, the vase was broken in transit. Although Mr John succeeded in cementing the pieces back together again, this irreplaceable English vase was ultimately destroyed in a fire that broke out in the Brownfield pottery on Saturday 3 March 1894. Its destruction was said to be one of the greatest losses in the history of ceramic art.

Casting Metals on Lace (1887)

In a century of constant change, when intrepid inventors and men of science were making revolutionary discoveries, the artists of the age were pushing the boundaries of their own field of study in a search for new and more imaginative ways of creating their masterpieces. In 1887 an American metallurgist, Mr Alexander Ewing Outerbridge (1850–1928), announced his new method of casting iron and other molten metals on to lace and other textile fabrics, as well as leaves, grasses, and flowers, creating metallic works of art that were as intricate and delicate as a spider's web.

Firstly, the lace, fabric or vegetable tissue was carbonised. This was achieved by placing the material inside a cast-iron box, the bottom of which was covered with a layer of powdered charcoal. Another layer of carbon dust was sprinkled over the top, and the box was sealed with a tight-fitting lid. The whole was then heated gradually in an oven to drive away any moisture, and the temperature was slowly raised until the escape of blue smoke from under the lid ceased. The heat was further increased until

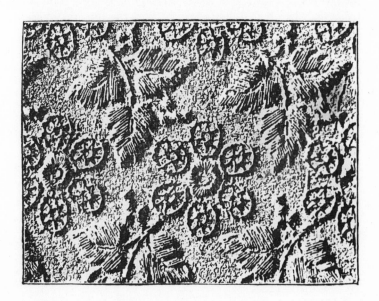

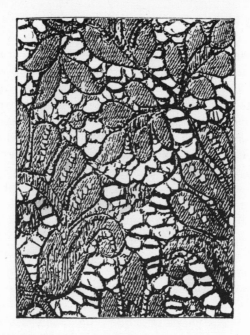

the box became white hot. It was kept in this glowing condition for at least two hours, then removed from the fire and allowed to go cold.

The carbonised lace was then spread smoothly on to a mould, and molten metal poured on top. This was left to cool before the iron castings were removed from the mould. Upon the face of the casting, Mr Outerbridge found a sharp and accurate reproduction of the original design. The lace had formed its pattern on the under surface of the molten metal, thus forming a die. This was used for a variety of purposes, such as embossing leather, stamping paper, or even producing pieces of jewellery.

The figures represent samples of castings as they were made.

Pictures in Stone (1884)

For many a long year a scientist named Dr Hand Smith had been engaged in perfecting his own artistic process, which constituted a brand-new departure in art. By 1884 Dr Smith had successfully produced pictures not on, but *beneath* the very surface of stone, marble, ivory, terracotta, and other solids. He had thereby, in effect, succeeded in combining the art of painting with the durability of sculpture, a result never before achieved in ancient or contemporary art. Though the secret to this technique was known only to Dr Smith, it was disclosed that by using metallic oxides as pigments, worked in a special medium and fixed by a classified treatment, the colours could be driven into and fixed within the stone, penetrating unchanged to any desired depth.

The process was said to be applicable to statuary, pottery, and architectural mouldings, and samples of Dr Smith's new art, which included decorative scrolls and delicate paintings of foliage and flowers, were exhibited during the 1884 Easter festivals at the seventeenth-century manor house in London, known locally as 'Piccadilly Hall'.

The Hoeschotype (1885)

Soon after Dr Smith had refined his stone picture process, an innovative method of accurately copying original paintings was developed. The technique was termed the 'hoeschotype', and the inventor used only five colours – yellow, red, blue, grey and black – to achieve his result. These hues, he explained, formed the basis of a large key map of tints. In combining these in differing quantities over 1,600 shades could be created.

The original painting was first photographed, and copies printed. One of these copies was taken in hand by an artist, who, by reference to his colour-scale, ascertained for each spot in the picture the exact amount of yellow it contained, and he covered that particular spot with an equivalent shade of yellow, at the same time painstakingly painting out with white all other parts of the print which would contain no yellow.

Once this laborious process was finished, a negative was produced from the painted sheet, and a print taken on sensitised gelatine mounted upon plate glass. This print therefore represented a picture of those parts in which only yellow appeared, and in different degrees of density.

Similarly, negatives were prepared for the four other key colours, and these were all printed one above another on one sheet of paper or canvas in perfect register, with the result that a very faithful copy of the original painting was obtained. Furthermore, it was possible by this method to make several copies of the picture.

The Airbrush (1887)

Many of the great Victorian artists, such as Edward Armitage (1817–1896) and Leeds-born John Atkinson Grimshaw (1836–1893), employed oil paints and tiny brushes to produce fine detail in their work. Since the time of the great Flemish painting masters of the fifteenth century, the chosen medium for pieces of artwork has traditionally been oil, as this type of paint was found to be viscous and slow to dry, allowing the artist to blend the paint together for a longer period until the desired effects had been produced. Photography, however, provided an emerging new way to capture a lifelike image, and with colour photography on the horizon, artists felt the pressure to invent more impressive ways in which to fashion their masterpieces, lest this new technology rendered their skills surplus to requirement.

Contrary to popular modern belief that the 'airbrush' is a twentieth-century digital retouching device, designed to generate flawless images of models and celebrities, this handy little tool was actually in use in Victorian times. The 'artistic brush', or 'colour applier', was patented in America in 1876, and received high praise from the Franklin Institute's Committee on Science and the Arts in Philadelphia.

The new instrument did not make up for any lack of artistic skill; it simply served to shorten the time required to apply tones of colour to works of art. As the airbrush threw the colour down into the pores of the paper or canvas, the various tints appeared equally well whether the light fell on them from one side or the other. This was not always the case with crayon colouring.

The nineteenth-century airbrush consisted of a spoon-like reservoir, which contained a little of the colour to be applied. Through this liquid, a fine needle darted rapidly backwards and forwards, its wetted point projecting each time beyond the edge of the spoon. A strong current of air from a pair of bellows, manipulated by the foot of the artist, blew past the needlepoint, and sprayed a delicate shower of paint against the paper. If the point of the needle was held close to the paper, the result was a fine line, which could be broadened by withdrawing the point.

If the Franklin Institute could have foreseen what a contentious manipulator the airbrush was to become over the centuries perhaps they would not have been so keen to give this little invention their stamp of approval.

4

EDUCATION, WORK AND INDUSTRY

The Diminutive Dictionary (1892)

Bryce's Thumb English Dictionary was published in 1891. This, however, was no ordinary publication, for this little Victorian book displayed a remarkable feat of condensation, measuring barely more than 2 inches by 1 inch, and was scarcely three-quarters of an inch thick. It was said, nevertheless, to contain a remarkable 15,000 references, and could duly serve as a guide to the spelling of many words in general use.

Mr David Bryce brought out several editions of his diminutive dictionary during the nineteenth century, including his *English Dictionary*, which measured 2½ cm by 2 cm. It came inside a metallic locket for transportation and protection purposes, complete with a magnifying lens in the front.

Only a handful of these dictionaries were ever printed, and in 2006 a copy of Mr Bryce's *Thumb English Dictionary* sold at Christie's for £840.

Small Writing (1893)

In France the curious fad of small writing had swept the nation by 1893, and *Le Petit Journal*, a popular Parisian daily newspaper, received from one of their readers an ordinary postcard that was said to contain more than 3,000 words. This inspired the rival journal, *l'Eclair*, to offer a prize to whoever could fit the most words into the smallest space.

This sudden trend of small writing was by no means a new one. Indeed, the Roman author Pliny the Elder (23–79 AD) described a copy of Homer's *Iliad* which had been written on a fragment of parchment that was contained inside a nutshell, from which we derive the familiar expression. Centuries later, a piece of paper measuring about 1 square cm which displayed the entire Ten Commandments was presented to Elisabeth of Austria, Queen of France (1554–1592), and in the 1890s, according to the *York Herald* newspaper, a Viennese gentleman, Herr Sofer, engraved a psalm of 391 letters on a single grain of wheat. Developments in the industrial printing press had led to similar triumphs. In 1889, at l'Exposition Universelle, there was exhibited a tiny copy of Dante, printed on pages measuring an inch square, but in setting the miniature type the compositor had strained his eyes to such an extent he almost went blind.

The Geodoscope (1892)

Reference books conveyed essential knowledge from one generation to the next, making them principal tools in the education of Victorian scholars. The written word, however, was only of any use to those who had learned how to read. Before 1880, the year when the education of children was made compulsory, there were many individuals, particularly among the working classes, who were illiterate.

In 1892 a young schoolmistress from Barnsley hit upon the idea of making education more accessible. She invented a visual tool of reference, one which was particularly useful as an aid to scientific instruction. Miss Annie Margaret Gregory L.L.A. (1857–1927) was employed in a girls' high school when she envisaged the 'geodoscope', a fully functioning miniature combination of the terrestrial and celestial globes, which she proudly exhibited in Chicago at the World's Columbian Exposition of 1893. The terrestrial globe was encompassed within a larger celestial globe

of glass, marked with the principal stars, the signs of the zodiac, the ecliptic, the celestial equator, and the tropics. The earth was rotated by means of a handle, H, projecting through this crystal shell, and all its phases were visible to the eye.

The earth was made from a 3-inch globe, the heavens a 12-inch. A cylindrical beam of light reflected from a hand lantern represented the sun, and by its rays falling on the earth it gave a perfect illustration of the day and night throughout the year.

The Allan Glen Revolver (1892)

In Glasgow a wealthy businessman named Mr Allan Glen (1772–1850) left a considerable sum of money upon his death for the

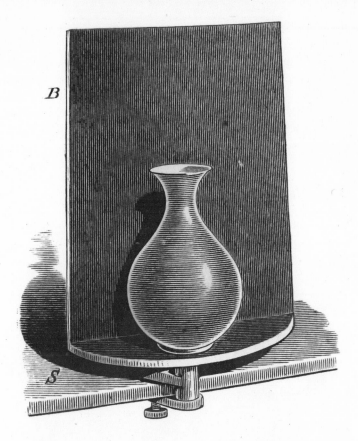

endowment of a new school in the city for the benefit of the sons of humble tradesmen. The institute was opened in 1853 and operated until 1989, when it was taken over by the College of Commerce.

By 1892 Mr John G. Kerr, the headmaster of Allan Glen's School, had introduced a new method of teaching perception and mental calculation to his pupils. The device he employed, called the 'Allan Glen Revolver', consisted of a revolving blackboard, B, having a shelf, S, attached, upon which an article could be placed. On the side of the board turned away from the pupils, rows of figures, dots, or letters were marked, and the board was spun round at different speeds. The pupils had to work hard to read the words or figures, and count the dots before they vanished from sight. The exercises were graduated, and became more difficult depending on the ability of the students.

The board was also used in freehand drawing classes to teach students the art of grasping and fixing contours in their minds. For example, a jar was placed on the revolving shelf, and the students were expected to draw its outline after it had vanished from sight.

Convertible School Furniture (1885)

In a world where comfort was not always a priority, particularly with regard to children and the lower classes, it was observed that the continual use of rigid desks in both schools and offices

tended to cause physical deformation to the sitter. With this in mind, a new manner of adjustable furniture, in the form of desks and chairs that could easily adapt to the ergonomic requirements of the user, was invented. They were so fitted and jointed that a person could effortlessly raise or lower either the desk or the seat of the chair to suit himself.

The Travelling Office Chair (1885)

An American inventor took pity on the poor Victorian bookkeepers and clerks whose routine duties compelled them to make frequent trips across the office to fetch and return one book after another. He devised for their use a special chair that ran through their office on a miniature tramway, in such a manner that a push of the foot upon a supplied footboard would propel the chair left or right across the room as desired. This enabled the worker to move from one book to another with greater ease, thereby reducing the unpleasant confusion that was often occasioned by the continual motion of the clerk in getting up and down from his seat all day.

The Portable Columbia Typewriter (1887)

For secretaries, reporters and others whose day-to-day tasks involved a considerable amount of writing, a portable typewriting machine that printed both capital and small case letters was invented. The 'Columbia', as it was called, was considered by Victorian typists to be a great improvement on its predecessor, which was deemed far too complicated for ordinary use. Additionally, the typewriter came in a polished case, which made it particularly suited to travellers and commuters.

The form of the Columbia may be understood by the illustration. On turning the handle, A, the indicator, B C, was revolved on the dial-plate, across the top of which the letters and figures were arranged in order. Unlike older typewriting machines, the letters were given only once, but the indicator was so arranged that when point B was directed to a letter, the type wheel, D, printed a capital. When the other point, C, was presented, a small letter was printed.

In using the apparatus, the required letter was found with the indicator by turning the handle to the right or left, and then depressing the handle: a process which automatically

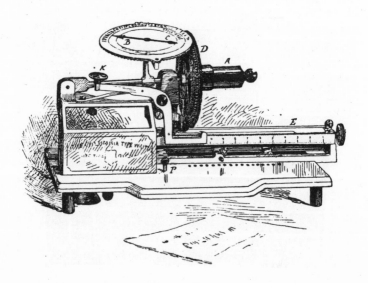

inked the letter and impressed it on the paper. Left to itself, the handle would at once recover its position, and the process could be repeated until the typing of the word was completed. The spacing of the words was implemented by the spacing key, K, which, when struck by the typist, moved the paper an appropriate distance.

The paper was carried in a carriage, E, which moved automatically from right to left as the writing proceeded, and was provided with a knob by means of which the paper was moved vertically, and the proper space between the lines ensured. By means of a little pin, P, the width of the line of writing was controlled, and, in order to secure a straight margin at each edge, an attachment was added which rang a little bell when the end of the line was reached.

The Columbia may have become obsolete over the centuries, but in 2006 one of these typewriters was sold at Christie's for £1,680 – more or less double the price of an 11-inch Apple MacBook Air laptop.

The Copying Apparatus (1887)

In 1887, around half a century before the arrival of the modern-day photocopier, a new kind of office apparatus for producing any number of duplications of a particular document was patented under the name of 'Eclipse'. The document to be copied was first written in prepared ink, and allowed to dry without blotting. It was then placed face downwards on the gelatine plate of the machine, and allowed to remain there for a few seconds until an impression of the writing was made. The negative thus left on the plate was inked by slowly passing a roller over it. A print on a new sheet of paper could thus be obtained from the inked gelatine, which required re-inking with the roller after each impression had been made.

The Businessman's Dinner Table (1884)

Despite these little improvements, days were often long and tiresome for the average Victorian businessman, and the more diligent worker was occasionally required to spend many a late hour sitting in his office. With this in mind, a new piece of furniture, which combined a table, a cooking stove, and a pantry, was made available to those who found it necessary to take their tea or other light meals in their offices.

The framework of this table was made rather deeper than usual, and came with a bottom, between which and the top of the table was a considerable space. The top of the table was detachable, so that when attached it formed an ordinary table to work or eat from, and when removed it revealed the secreted compartment. The centre of this was occupied by the stove, which could burn either gas or oil, and was useful for boiling a kettle or cooking food. Round the lamp was enough room to store the various articles of tableware, which would ordinarily be

hidden from view when the table top was replaced. Between the legs of the table stood a useful box which served to keep bread and other provisions inside; thus the businessman could stay in his office as long as was required without going hungry.

The Office Indicator (1893)

For those times when professional men did not wish to be disturbed, an inventor named Mr Jenkins patented a new type of office indicator, which was in use in Britain by 1893. The dial, which was fixed outside the door, displayed a clock face with an opening at the top, through which could be espied one of several self-explanatory announcements carried on a revolving dial, controlled from inside the room by a very simple mechanism. The indicator could thus be made to show that the tenant of the room 'will return' at any particular hour, or that he was 'within', was 'engaged', was 'out of town', or that it was 'mail day', and was therefore too busy to be disturbed by trivial enquiries.

The Indicating Door Bolt (1887)

The illustrations show a similar type of door bolt, one that proved invaluable to members of the public wishing to secure themselves against embarrassing interruption. On being pushed home, the clever device displayed the notice 'Engaged' outside the lavatory door, as shown overleaf. When the bolt was withdrawn, a metal-plated shield took the place of this word. The top diagram shows the bolt on the inner side of the door, and the ratchet and wheel by which the movement of the bolt worked the indicator.

This modesty-saving invention was the brainchild of Mr Arthur Ashwell (1829–1903) of Herne Hill, London, who was sitting on a train to Waterloo one day when he was troubled by an urgent need to answer a call of nature. Too embarrassed to try the door of the nearest convenience, he suddenly conceived an idea that would become known as 'Ashwell's Door Indicator Fastener'

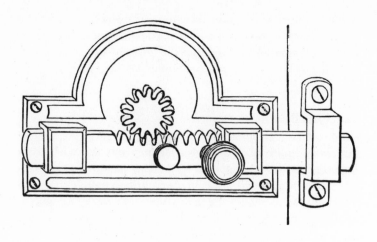

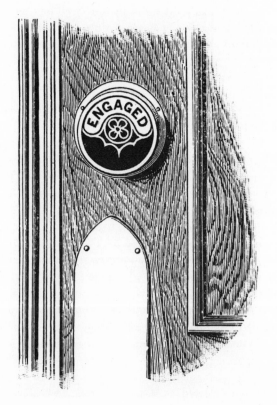

and would eventually make its inventor positively flush. When Ashwell died in 1903 his estate was valued at £13,978 4*s* 9*d*: an amount in excess of £1 million in today's money.

The Checking-In Clock (1885)

While office door indicators were handy devices for quickly determining the availability of the management, a new checking-in clock to automatically record the attendance of a firm's

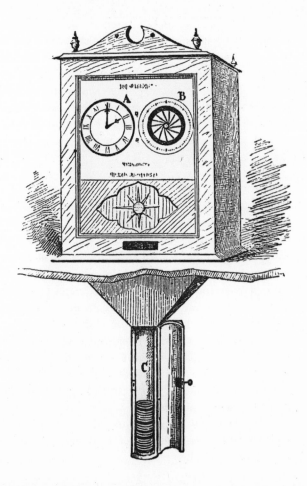

employees was introduced in 1885, and is thought to be one of the first of its kind. It consisted, as shown in the figure, of an ordinary clock, A; a slotted disc, B; and a tube, C. The object of the latter was to collect the personalised checks as the workmen deposited them into a narrow slit below the clock. The checks were piled up in the tube in the order in which they were dropped, thus revealing the order of arrival of the workmen. If an employee should misleadingly insert the check for an absent comrade, whose nonattendance was later discovered, the dishonest workman could easily be found out.

Inside the disc, B, were a certain number of slots (for instance, twelve), each containing a copper time-check, and the disc could be set to revolve at a specific speed, say one revolution per hour. Below the disc was a passage communicating with the check-tube, C, and as the disc revolved, the copper time-checks would fall one by one, landing in the check-tube among the workmen's checks. If the checking-in clock was set so that one copper time-check fell every five minutes, by its position among the workmen's checks, it revealed the time within a few minutes at which the workmen arrived.

The Watchman's Tell-Tale (1887 & 1885)

A similar plan for checking the fidelity of night watchmen while on patrol proved to be a successful introduction at factories, warehouses, and other places of work. The invention consisted of a series of posts or boxes arranged in an electric circuit. These posts were placed at convenient points on the watchman's beat, and each comprised a keyhole into which the watchman inserted a key as he went along. This switched on an electric connection inside the post between the wire coming to it and the wire leaving it. When the last post on his beat was keyed, the circuit was completed, and the current from a battery would flow through all the posts. This actuated a pencil in the manager's office, which mechanically recorded on a strip of paper rotated by clockwork that the watchman's beat had ended, and also the time at which this had taken place.

An alternative tell-tale was invented to record the reliability of the night attendants on duty in Britain's asylums – institutions which had fast increased in number since 1845. This device was the design of Glasgow-born Dr John Millar (c. 1819–1888), the medical superintendent of the Bethnal House Asylum at Bethnal Green. Though Bethnal House Asylum serves as a public library today, in 1881 it was home to well over 100 female inmates of all ages, and from varying backgrounds. Their number included a prostitute, a lady from Devonshire, and even a clergyman's wife.

The night watchman's tell-tale consisted of a clock movement, so arranged that the hour axle rotated a paper disc instead of the usual hand. This was encased in a strong mahogany box, which was kept locked, and a number of these apparatuses were fixed at convenient points on the beat of the attendant. In passing a tell-tale box, he inserted a pencil as far as it would go into a hole in the lid of the box, thus marking the paper disc, which had previously been set by the management so as to correspond with the time at which each mark was due to be made. When the clock was examined the following morning, the record would reveal whether or not the watchman had kept to his beat during the night.

The Ice Harvester (1884)

Industry was flourishing across the Western world during the nineteenth century. In America ice harvesting was one of the most productive businesses, and the total annual ice-yield of the States was reportedly 20 million tons, of which some 12 million was intended for domestic consumption. Mining and storing this ice had given rise to a separate branch of engineering altogether, and special implements for the purpose were required. Various kinds of scrapers to remove the snow were invented, as were tracers, or hand-ploughs, to mark out the areas to be cut. A tool called a marker was fitted with knife edges, which, on being lowered to the grooves in the ice, cut straight through it.

Once the ice had been cut and cross-cut with these tools, it was sawn by the ice-plough shown in the following image, until

two-thirds of the total thickness had been cut through. This plough consisted of a succession of curved blade-like teeth attached to a long beam. The teeth were so formed as to clear themselves, and to carry the chips out of the groove with little resistance.

A channel was cut by the above means between the ice field and the elevators that raised the excavated blocks into specially constructed ice houses. The blocks were loosened from the ground by ice chisels, floated to the elevators, and raised by steam power on endless chains working up an inclined plane.

The ice rooms were built at 100 feet long by 40 feet wide, and provided with suitable insulation, meaning that the ice would remain frozen for many months, so that in the summer a ready supply of natural ice was available to be sold and shipped anywhere in the world.

The cost of all this preparation was calculated to be just 25 cents per ton, and in the days before modern refrigerators, America's ice harvesting industry helped to revolutionise the way in which meat, fruit, fish and vegetables were stored and transported. Merchants of such perishables began to make more money than ever before, as it had become possible to tap into an international market by delivering fresh produce directly to the consumer, who would have hitherto purchased their meat and fish from local dealers. Using specially adapted refrigeration cars, ships, and railway carriages, the food could be transported over hundreds of miles. This also meant that fishermen were able to travel further, into more abundant waters, in order to catch the biggest and best fish without the fear of them perishing during a long journey home.

The Grain Dryer (1884)

While the American frozen food industry was prospering, back in Britain Victorian farmers had stumbled across a rather watery problem. Owing to the country's habitually moist climate they often encountered great difficulty when attempting to dry their produce. It was a gentleman named Mr Gibbs who offered an excellent remedy to this soggy setback. His invention was said to be capable of drying from 10,000 to 14,000 bushels of grain per week, and was, by 1884, in regular use on the estates of the Earl of Eldon and other landowners in Hereford, Paisley, and Londonderry.

The figure below illustrates the invention, namely, a 3-to-4-ton machine for drying grain, malt, hops, cider pulp, rolled tea, and other damp produce. It consisted of a metal cylinder, E E, which communicated at one end with a drying fan, B, and was supported in an inclined position by brick piers. The gear shown at F was for raising and lowering the cylinder, while the pier, A, was either a furnace in which coke was burned, or a

flue conveying waste heat from some other source. A chimney,
C, carried away the smoke from the fire, and a thermometer, *c*,
showed the temperature of the hot air as it passed to the grain.
This air was let through the cylinder by ducts shown at G. The
grain was fed into a hopper, D, and after being carried through
the cylinder in revolving cells, H, it was poured out by a shoot at
the other end in a continual stream, providing the farmer with a
perfectly dry supply of grain.

The Tea Dryer (1884)

Another contraption for drying moist tealeaves was invented
by Mr J. Greig of Edinburgh, and was found particularly useful
in Ceylon and India. The machine consisted of a cylindrical
hollow receptacle with wire-cloth sides, which contained the
damp produce. A supply of dry air was warmed to any desired
temperature, and then pumped by the flue, as shown, into this
cylinder, where it circulated through every part of the tea. The
belts and pulleys caused the cylinder to revolve, thereby ensuring
a thorough drying, and the resulting steam escaped through the
wire-cloth.

The Exploding Scarecrow (1893)

Throughout the ages, farmers have been faced with the perpetual problem of how to prevent their crops from being consumed by birds and vermin. Even the traditional scarecrow, after a time, became less of a concern for the local wildlife owing to the fact it was immobile. In 1893 a Kentish gentlemen came up with a rather drastic way to frighten away the birds in the form of an exploding scarecrow. This radical invention consisted of a hollow post, as shown, comprising a row of twenty-four holes (H H). Each hole contained a cartridge, and in the centre of the post was a slow-burning fuse, M, which zigzagged from one cartridge to the next, firing them in turn. The post was perforated with further holes to allow the sound to escape, and the plumes of smoke rose through an opening at the top. A cowl was fitted over the pinnacle of the scarecrow to keep out the rain. The fuse could be timed to start the explosions at daybreak, the traditional feeding time of birds who were keen to catch the early worm.

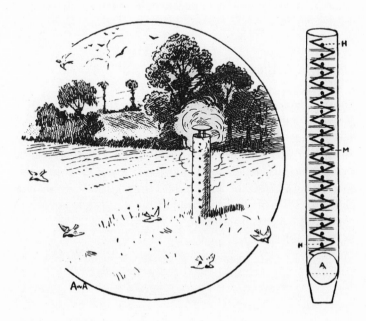

The Potato Digger and Picker (1884)

The automatic potato gatherer was yet another example of how new technology was helping Victorian farmers carry out their agricultural duties. In this new machine two ploughs were arranged so as to part the soil of a farmer's field on either side, while a third, single plough passed beneath, raising the potatoes and the earth in which they were lodged. A shaker, or picker, separated the tubers from the soil and delivered them to the rear of the machine for collection, the mould falling through the picker back to the ground. With this new invention, a whole potato field could easily be cleared in a tolerably short space of time, whereas this task would have hitherto taken many hours of physical toil.

Though much hype surrounded these automatic machines, Victorian technophobes dismissed such contraptions as nothing more than whimsical fads, holding the belief that the suspension of manual labour would not last long. Over time, however, it became apparent that the sceptics were mistaken, for as the months and years went by more and more agricultural machines were invented.

The Fruit Gatherer (1892)

The new 'fruit gatherer' was devised to effortlessly facilitate the gathering of delicate fruit, so situated in the tree as to place them out of reach by means of a ladder. The brass cups on the head of the picker, resembling a small pair of cymbals, were opened by a cord which passed down the long handle, and as soon as this cord was slackened the India rubber ring just below the cups pulled them together, and so gripped the fruit gently enough to prevent damage or bruising.

The Steam Man (1893)

Professor George Moore was a Canadian inventor who dreamed that one day mechanical robots would relieve mankind of every laborious task imaginable. Before the turn of the century he began constructing one of the most ambitious and revolutionary industrial developments of the age. Gripped by his idealistic notion, he set about devising a mechanical man powered entirely by steam with the purpose of pulling either a cart across a farmer's field or a waggon containing up to ten commuters through the city streets.

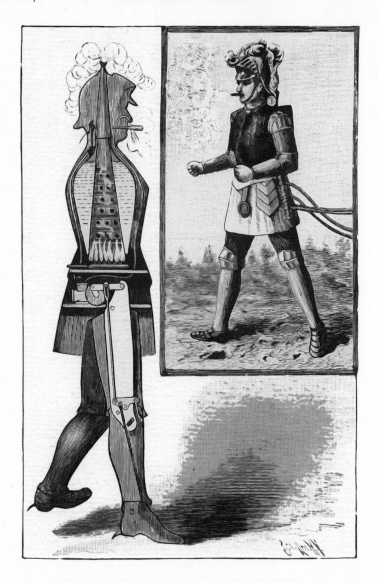

Once completed, this futuristic figure was said to stand 6 feet tall, and could walk at a brisk pace of 5 miles an hour. The 'man' was made of steel and tin, and fashioned in the likeness of a medieval knight in armour, as depicted in the image taken from

Mr Albert A. Hopkins's 1897 publication, *Magic: Stage Illusions and Scientific Diversions*. The top of the smoke flue, from which the by-products of combustion escaped, was concealed by the plumes of his helmet, and an engine exhaust pipe, which led from the engine to his visor, was disguised as a cigar. The trunk of the body contained the furnace, boiler and engine, and the limbs the mechanisms for walking. So competently designed were its inner workings that the figure's hip, knee and ankle movements were described by alleged witnesses as being remarkably lifelike.

The fervent professor supposedly exhibited his incredible feat of engineering widely across America, though the ultimate fate of the steam man remains a mystery. Some claim to have found its iron relics in a New York junk store. Others remain unconvinced that it ever existed in the first place. Whatever became of this nineteenth-century robot, it was claimed that, when set to full power, two fully-grown men would not be able hold it back. Perhaps Moore eventually gave up hope and abandoned his ambitious project, or perhaps, in the end, the steam-powered android walked out on the professor.

The Steam Tree-Feller (1892)

Back in Britain, a new kind of steam-powered machine, as shown in the engraving overleaf, was making strides of its own. The combined tree-feller and crosscut saw was designed for operation in the middle of the woods, and could be fixed to a tree in just a couple of minutes. It was said to be capable of automatically felling or sawing trees up to 4 feet in diameter, and, as it was only quite light, could be carried about by four men. In five minutes it could saw down an oak tree 3 feet in diameter, and eight trees, averaging 30 inches in diameter, in an hour. It worked in any position, and could be 'readily applied by any man of ordinary intelligence following the printed instructions'. The steam was conveyed to the motor by means of a flexible pipe from the generating boiler, as shown in the figure.

The Firewood Machine (1887)

Once the trees had been felled, they were fed into a machine that was designed to cut up old timber into firewood using an automated chopping knife, before automatically tying the wood into bundles ready for sale. The timber was slotted into the machine in lengths of about 6 inches, and the knife split them up along the grain of the wood. Next they passed out of the machine into a box, which by agitation settled them into their smallest bulk. They were then formed into bundles by mechanical means, compressed, and bound by wire.

The Automatic Cotton-Picker (1885)

Cotton-picking was an activity that for many centuries was carried out by hand, but by 1885 an American inventor had introduced the first mechanical picker. This contraption weighed

800 lb, and was capable of picking four bales per day. In other words, one machine could do the work of thirty men. It was mounted on wheels so as to bestride a cotton row, and was designed to glean each side of the row. The picker plates were set with teeth, which traversed the whole of the cotton plant where there was likely to be any cotton, and stripped off the wool while allowing the branches and leaves to pass through. The cotton was caught between the teeth, which were cleared by means of rapidly revolving brushes. The cotton was then sent into a receptacle, ready for collection by the machine operator.

The Sheep-Shearing Machine (1893)

The production of wool was another prominent industry in the nineteenth century, and a new type of machine for shearing sheep is illustrated overleaf. The principal novelty of this new device was its flexible driving shaft of steel rods connected by toothed gearing. The cutting teeth of the shears derived their motion from this shaft through a universal joint, which allowed the shears to be turned in any direction, rendering the job

quicker and easier than when using manual shears. This new machine required just two operatives – one to turn the driving handle and one to operate the shears – and it was capable of sheering the fleece off a sheep in three minutes.

When not in use, the implement could be readily taken apart for cleaning, transportation or storage.

The Rug Machine (1885)

Once the wool had been collected, farmers would distribute their produce to the imposing textile mills and factories across the country to be transformed into a variety of products, such as clothes, blankets, shawls, or rugs.

For home use, however, an ingenious machine was invented which automatically made rugs from wool, cotton, yarn, unwanted rags, and even old clothes. This domestic rug-making machine, which could be easily operated by any member of the family, was constructed from steel, and was quite capable of producing a rug in just a few hours.

Firstly, the operator was required to form a foundation from some strips of old fabric, before feeding the material or waste clothes into the weaving machine. These were then automatically woven around the formation, and a brand new rug, fashioned from the family's unwanted scraps, would thus be ready to lay by the hearth for all to admire.

Uses for Asbestos (1884 & 1885)

In the cloth-dyeing industry it was often necessary for the textile workers to hang their pieces of fabric in loops from parallel rods, in order to expose them to the action of steam, air or ammonia. This was usually done by means of ropes, but the toxic vapours in the factory atmosphere quickly rotted these away. Asbestos binders were therefore employed in place of the ropes, reportedly with great success, as the corrosive fumes had no influence upon them whatsoever.

Other promoters of this fibrous mineral were Herr Erichsen of Copenhagen, and Mr Toope of England. The former gentleman discovered that by using powdered asbestos to make an enamel for coating pipes and walls, the surfaces became resistant to the corrosive action of the elements, while at the International Health Exhibition of 1884, Mr Toope unveiled his own original application of the fibre. He had fashioned a safe made from asbestos, formed into a solid stone-like material. Such boxes, he claimed, were fireproof, soundproof, thief-proof, and vermin-proof, and therefore made excellent storage compartments for important documents. Despite their massive size, the safes were exceedingly light, rendering construction simple.

The Portable Aluminium Army Canteen (1892)

Another light and durable building material that was popular during the late nineteenth century was aluminium. Though ancient civilisations were aware of this chemical element, an industrial process for producing the metal was not developed until the 1880s, after which time it was used regularly in the global construction industry. In fact, the remarkable lightness of aluminium metal induced the German emperor, William I, to decree its use for the construction of portable canteens for his army. He theorised that such canteens would be easy to disassemble, carry over long distances, and reassemble again at the next station.

Upon closer examination, however, two German chemists, Herren Lübbert and Roscher, observed that aluminium was to a certain extent dissolvable in boiling water, and was therefore found unsuitable for use in the kitchen. Preserves, pickles, and acids in the presence of table salt also attacked the metal, as well as acetic, citric, and tartaric acids, Bordeaux and Moselle wines, and even infusions of tea and coffee. Hence the Kaiser's proposal for its use in the making of mobile Army canteens was ultimately a non-starter.

Portable Railway Bridges (1885)

The construction of other kinds of portable structures began during the latter part of the nineteenth century. This was due in part to the development of the rail network, which meant that

structures could be built off-site in specially prepared workshops and transported wholly to their intended destination instead of being constructed at their proposed location, thus saving time and money. By the mid-1880s this method had been applied to the construction of colossal railway bridges.

The illustration represents the mode of moving fully constructed bridges by rail. In this way bridges could be moved across any distance, supported on trucks which were drawn by locomotives, at a rate of 15 miles an hour. The bridge was then lifted into position by cranes.

The Forth Bridge (1884)

One of the most famous cantilever railway bridges of the late Victorian era was Scotland's Forth Bridge. This great construction was erected over the Firth of Forth, between North and South Queensferry, on the Fife and Lothian shores. Work was begun on this marvellous construction in 1883, and by the following year it was reported that a labouring staff of between 900 and 1,200 men was engaged, working on both shores as well as on the intermediate island of Inchgarvie.

The cantilevers sprung from four stone piers, out of immense caissons or hollow towers 61½ feet in diameter, sunk into the bed of the estuary and filled in with concrete. Each cantilever rose to a height of 350 feet above the piers, and stretched out an arm 650 feet long left and right of the centre, supporting the framework of iron. They were made from steel tubes, constructed at workshops at South Queensferry. So large were the tubes composing the cantilevers (being 150 feet long and 12 feet in diameter) that as they were built up, the whole workshop was shifted along them.

After seven years in the making the bridge was finally opened on 4 March 1890. It took 130 million lbs of metal to construct the bridge, 6½ million rivets, and ten times the amount of steel as the Eiffel Tower. It also cost the lives of as many as ninety-eight construction workers.

The Forth Bridge sported the largest span of a single cantilever in the entire world, a record that remained unbeaten until 1919, following the opening of the Quebec Bridge in Canada.

The Tower Bridge of London (1892)

On 21 June 1886 the Corporation of the City of London began construction of the remarkable structure known as Tower Bridge, which was expected to be completed in 1893. The architect was

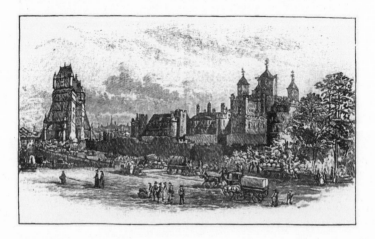

Sir Horace Jones (1819–1887), and the chief engineer was Sir John Wolfe Barry (1836–1918).

The River Thames was 880 feet wide at the point where the bridge was to cross. Once erected, it consisted of three spans – a central span of 200 feet, and two side spans of 270 feet each, which, together with piers and approaches, brought its total length to 2,640 feet. The shore spans were devised on the suspension principle, and the central span consisted of iron girders. The piers were built of grey granite, and the upper parts of the middle towers were of red brick with stone facings.

To allow for ships travelling up and down the river, the central span was built on the drawbridge principle – the two halves of the spans being hinged at the piers, capable of being raised into a vertical position and then lowered into the horizontal position after each vessel had passed through, causing only a short interruption to the vehicular traffic. It was expected that about two dozen ships would pass through the bridge on a daily basis, each taking no more than five minutes to pass through. The working of the drawbridge was effected by hydraulic power and all the engines were duplicated in case of a breakdown.

The foot passengers crossed the river by a lighter bridge, raised to a height of 135 feet above the river on the lofty middle piers.

The Prince of Wales, who eight years earlier had laid the foundation stone, officially opened the bridge on 30 June 1894.

The Niagara River Cantilever Bridge (1884)

The accompanying illustration represents the handsome railway bridge that was completed across the famous Niagara River. It was built to connect the New York Central and Michigan Central railway systems. The view shows the structure as seen from the American bank of the river, looking towards the celebrated falls, just visible in the background. This type of bridge was self-supporting during construction: an essential part of the design, as it was found to be impossible to raise temporary supports in the tremendous rapids below.

It was officially opened in 1883, and was situated some 300

feet above the old railway suspension bridge, spanning a gulf 870 feet wide from shore to shore. The bridge consisted of two huge steel towers, each resting on stone piers. The two towers supported two cantilevers, the shore ends of which were anchored to piers of substantial masonry. The total length of the bridge, upon which a double line of rails was laid, was 910 feet 5 inches between the centres of the anchorage piers, and from the surface of the water to the base of the rails the height was 239 feet.

After more than forty years transporting trains across the river, the Niagara cantilever bridge was eventually closed in 1925, giving way to a new railway bridge, which stands in disrepair today.

Niagara Falls Hydropower (1885 & 1887)

For centuries visitors to Niagara's magnificent falls have been held in awe by the sheer majesty of the spectacle, but some nineteenth-century observers expressed regret that the power of the water was not being utilised, causing one tourist to declare

that such a waste filled him with 'a grief altogether too deep for tears'. A statistical analysis of the potential power was published so that others could see for themselves the extent of the loss. Taking the height of the falls to be 150 feet, it was estimated that 1.165 billion cubic feet of water fell every hour. Mathematicians concluded that the power of the falls was thus equivalent to 5 million horsepower, or, in other words, nearly a quarter of the whole steam power of the earth.

The use of Niagara's waterpower for generating electricity in order to supply power to the neighbouring district was, to an extent, eventually realised by 1885 when a dynamo, and a mill to house it, was built beside the falls. This dynamo supplied the electricity for telephonic purposes to some 300 towns and villages, one of these being Buffalo, some 25 miles away.

Meanwhile, over in Scotland, experiments to light the town of Greenock using the water supply from the high hills in the district were also well underway. A portion of the town was lit by incandescent lamps, fed by a dynamo that was driven by a turbine at the waterworks. Though this experiment ran for just two years, it showed great potential, and in the Highland village of Fort Augustus, the resident Benedictine monks harnessed their own supply of hydropower to light up their abbey on the banks of Loch Ness, following the installation of an 18-kw turbine. It was never confirmed, however, whether this facilitated any sightings of the elusive monster during the newly illuminated twilight hours.

The Sewer Gas Cremator (1884)

Following advancements brought about by the Industrial Revolution, water began to play an important role in local communities, and not just for the production of electricity. Improved plumbing systems were slowly being introduced to the towns, cities, and even private homes, and the construction of Britain's modern sewerage infrastructure was implemented in London during Queen Victoria's reign. This subterranean labyrinth of pipes and tunnels flushed the raw sewage that was hitherto dumped in rivers, and even on the streets, deep under the ground. While this led to improved sanitary conditions for the inhabitants of the city, little did they know that under their feet a toxic and highly flammable build-up of deadly methane gas was rapidly increasing as more and more waste was pumped into the sewers.

At London's International Health Exhibition of 1884, an inventor named Mr Samuel C. Dean exhibited a 'sewer gas cremator', which he'd dubbed the 'parcaean'. This invention, which supposedly cleansed the air from the sewer before it contaminated the atmosphere at street level, consisted of a light sheet-iron or earthenware cylindrical casting, the upper part of which was filled with that dexterous material, asbestos. The whole was fitted inside the sewer, just below the pavement, and kept at a temperature of 500° Fahrenheit by a gas jet. The air that escaped from the sewer was forced to pass through the heated asbestos, which allegedly cremated the infectious germs and other hazardous vapours.

Notwithstanding the fact that large quantities of methane gas are potentially explosive, this new apparatus proved to be such a hit that it was also adopted as a ventilator on the roofs of hospitals, and also in chemical works for destroying germs and effluvia. It was said that one parcaean measuring 2 feet high and 9 inches in diameter would disinfect over 20,000 cubic feet of air in a day, helping to bring fresh air back to the smoggy cities.

The Glass Sewer (1884)

It was suggested that considerable sanitary advantages would result from completely replacing the existing sewers with ones made entirely from glass. There was little doubt among experts that such sewers would not only be very durable and easy to clean, but, owing to their smooth, hard surface offering no hold for the lodgement of refuse matter, would also possess great hygienic value. Unfortunately, the major drawback of this proposition was that glass sewers could not be constructed at a sufficiently low price, and as the London sewerage system had only been installed a couple of decades ago, the idea was ultimately discarded.

The Glass Floor (1884)

In Paris, glass flooring was slowly beginning to replace wood in many high-street shops. Though the initial cost was far greater than traditional floorboards, the glass lasted considerably longer and offered a more pleasing aesthetic effect.

By 1884 the whole of the ground floor of the headquarters of the Crédit Lyonnais bank on the Boulevard des Italiens was paved with large squares of roughened glass embedded in strong iron frames. This meant that in the cellars beneath there was enough natural light for the clerks to work by, and though it was not reported what passing ladies, in their long flowing dresses, thought of this plan, a great deal of money was saved on artificial lights and candles.

The central hall of the Comptoir d'Escompte bank was also similarly provided. The glass was cast in slabs 18 inches square and 1½ inches thick, and the rays of light the floor permitted were of a striking bluish tinge.

Glass Paper and Pulleys (1887 & 1884)

During the 1880s an increasing number of uses for this smashing material were dreamed up. Deadly as it may sound, powdered glass was deemed to be a far more suitable alternative to sand for lining sandpaper, and in Pittsburg, Pennsylvania, a firm of glassworkers was even manufacturing industrial glass pulleys. These were to replace the more traditional iron ones, which had a much higher coefficient of friction, and wore the ropes and belts out faster. The new pulleys, made from extra strong glass, were cast in a mould, and the annealing process took seventy-two hours. They were about 13 inches in diameter, and between 2½ and 3 inches wide, with a groove in the centre to receive the cable. The glass was found to be suitable for cable tramways, and though a little more expensive at first, were believed to be more economical in the long run.

The Postage Stamp Seller (1892)

While such developments were making industrial tasks simpler for manual labourers, the general public were also demanding greater conveniences for their everyday use. One such amenity that was met by public approval was an automatic stamp-selling machine, and by 1892 a prototype of this new device was tested by a private advertising company. Following the trial's success, it was hoped that the Post Office would adopt this machine as one of public necessity and affix such machines to pillar boxes, or the outside of their offices, unaccompanied by advertisements.

There were various designs for these machines, one of which is illustrated in the figure, where A was the slot into which the coin was dropped, and B was the stamp, delivered to the purchaser by the opening of the drawer into which it fell. It was felt, however, that this particular design had several flaws, and a later variant

of the machine dispensed the stamp through a slot, so that the purchaser only had to tear it off by the perforations instead of operating a drawer.

The Shop Weigher (1892)

By 1892 Señor Joce Solar, of Barcelona, had brought out another high-street convenience in the form of a new shop-counter balance, which was capable of weighing both large and small articles. This was accomplished by a pair of different-sized scale pans, one very large, the other very small, which balanced on the steelyard principle. In addition, this ingenious balance also checked the price of items by means of a third pan that moved along the steelyard arm, which was graduated in monetary values. The rate per kilogram being known, the money pan was moved to the corresponding position on the arm and money put in it until it balanced the weight of the article. Therefore the amount of money put into the pan equalled the price of the article.

The Automatic Piggy Bank (1892)

Money was often tight for many of the poorer Victorian families, so automatic 'piggy' banks, which only opened when a pre-arranged amount had been collected in them, were found extremely beneficial. One such model is shown in the engravings. The figure below gives an idea of the outside, and the figure opposite the interior.

If a person wished the bank to open when they had put, say, nine sixpences, or the same number of half-sovereigns, into it, the spring, A (shown opposite), was lifted, and the numbered disc, B, was turned to the right until the arrow pointed to nine. The lid was then locked, and the coins to be saved were placed in the circle, C (shown below). When the pin, D, was pushed to the left, the coin was admitted into the box, and automatically registered. Each time a coin was received in this way, the number shown at C became less by one, thus informing the depositor how many more were required before the sum was complete, and the whole could be drawn.

Such a contrivance, which only cost a few pence to purchase, was not only useful to children who saved pocket money, but

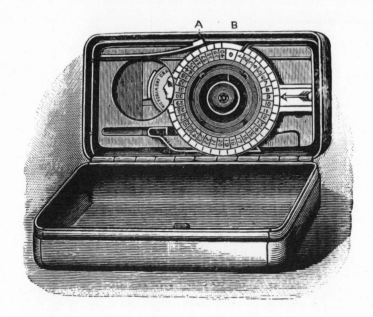

also to adults who were required to keep a close watch on their expenditure. These banks were also helpful to families desirous of saving enough money to fund the purchase of something special, such as the latest electrical gadget.

5

ELECTRICAL INNOVATIONS

The Influence Machine (1893)

In the early nineteenth century some of the greatest scientific minds of the era, such as Messrs Michael Faraday (1791–1867) and George Ohm (1789–1854), conspired to develop mankind's understanding of electricity. Their remarkable progress laid the foundations for the next generation of physicists, including Messrs Thomas Edison (1847–1931) and Nikola Tesla (1856–1943), to begin perfecting the art of electrical engineering. These distinguished scientists reasoned that if this mighty source of energy could be harnessed, it could be used to power countless new machines and contraptions, lighting the way into a bright new century.

Mr James Wimshurst (1832–1903) was one such electrical engineer. By the early 1880s he had invented his own version of the 'influence machine', or electrostatic generator, for generating electricity by induction. His invention was capable of producing a far higher voltage than the old friction machines of the same size could yield, and by 1893 he had unveiled an upgraded design. Mr Wimshurst's new machine consisted of two plate-glass discs, D D, each 3 feet 5 inches in diameter and a quarter of an inch in thickness, mounted on one boss and spindle less than an inch apart. When the spindle was turned by the handle, H, the discs rotated in the same direction, and by means of a series of paper

inductors mounted on crescents of glass, C C, between the discs and fine wire brushes, B B, for collecting the charges, a copious supply of electricity was obtained, which led to the discharging points, P. It was hoped that such an electrical discharge could one day be made strong enough to operate household appliances.

The Lightning Guard (1892 & 1884)

Though electrical engineering was still in its infancy by the late Victorian era, the introduction of electricity as a source of power

had slowly begun to weave its way into everyday life. Power cables had been installed in some of the bigger towns and cities across the world, and while this offered immeasurable advantages to its residents, there were also many dangers associated with these untested electrical infrastructures.

In 1884 a terrible thunderstorm in Virginia passed overhead just as tour guides were showing a party through the famous Luray Caverns, which had been discovered by local explorers just six years earlier. A great bolt of lightning struck the electric light wires and entered the cave, shattering the newly installed electric lamps, which exploded into hundreds of glistening pieces among the beautiful stalactites of the 'Bridal Chamber'. The scene was like a pyrotechnic display, and the party was rooted to the spot in terror. When the discharge dispersed, they were left in total darkness and found it almost impossible to retrace their steps to the exit.

By 1892, the British physicist and inventor Sir Oliver Lodge F.R.S. (1851–1940) had invented a little apparatus to protect electric light installations from being hit by lightning. It is illustrated in the figure, and consisted of a cylindrical barrel, B, through which the current passed by the axis, A, and two coils of thick wire, W W. At equal intervals in the circuit of this wire were three pairs of brass collars, C C C, with squares of sheet mica, coated with tin foil, between them. Mica was selected for this purpose owing to its insulation properties. The corners of the foils came close to the metal barrel or case of the apparatus, which was connected to the earth. In the case of lightning entering the coil, the electricity would leak away or branch off by the tin foils and, in theory, escape safely to the ground.

Electric Power for Paris (1885)

With such measures being installed to guard against danger, the French electrical engineer Monsieur Marcel Deprez (1843–1918) arranged to transmit from 100 to 150 horsepower of energy from Creil in the north of France to the workshops of the Gare du Nord railway station, Paris, in order to power the machinery. This was a distance of approximately 30 miles. As it was calculated that about 50 per cent of the power would be lost in transmission, Monsieur Deprez was required to generate from 200 to 300 horsepower at Creil. This, he discovered, could be provided cleanly – and without cost – by waterpower.

The apparatus he used consisted of two powerful dynamos, one in Creil and one at the receiving end in Paris, which were connected by a conductor that ran between the two. Turbines at Creil transformed the waterpower into mechanical rotation, which drove the generating dynamo. This in turn transmitted the energy in the form of an electrical current to the other dynamo at Paris, which reproduced the energy as mechanical power to operate the machines in the workshops.

The Electric Gun (1887 & 1884)

Towards the end of the century it was not just towns, cities and public buildings that were being fitted with electrical power. For some time, the War Office had been engaged in testing this modern source of energy in the working of weapons, and by 1887 it had been announced that the guns at Spithead Fort, Portsmouth, were to be fitted up with new electrical training gear, powered by a huge generator.

Over in Europe, the notable German gun maker Herr Henri Pieper (*c.* 1840–1898) founded a weapon-manufacturing company in Belgium. One of his weapons was a new type of electric gun, in which the powder was exploded by a platinum wire, heated by an electric current that was supplied by a small accumulator carried on the soldier's belt. The gun was capable of firing 10,000 rounds, which was considered a great achievement for its day.

The arrangement was such that it was the soldier who was rigged up in an electrical circuit. One pole of the accumulator was connected to a metallic glove, worn on the left hand, and the other to a metallic shoulder strap. When the gun was brought to the shoulder to take aim, the circuit was almost complete.

The figure below shows a section of the cartridge. M represents the gunpowder chamber, the axis of which was occupied by the copper rod, a, which applied pressure on the bottom plate, but from which it was insulated by a small ebonite piece, b. The rod, a, pressed on a brass button, c, of the disc, d. Close to this was the platinum wire, f, a tenth of a millimetre in diameter, connected on one side to the button, c, and on the other to a strip,

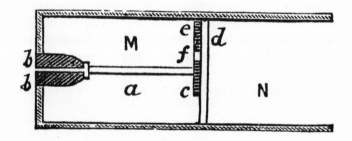

e, attached to the cartridge shell. As soon as the soldier pulled the trigger, the firing pin was pressed against the end of the rod, *a*, and the circuit was closed. The platinum wire was heated and the charge exploded.

Herr Pieper's new gun was exhibited at the International Electrical Exhibition, which was held in Vienna in 1883, where it attracted considerable notice from all who flocked to see the latest electrical inventions of the day.

The Magazine Gun (1887)

While electricity was helping to power innovative firearms, traditional guns were still being experimented with, and by 1887 a new type of repeating rifle, which could also be used as an ordinary breech-loader, was tested by the government authorities at Enfield. This magazine gun was the invention of the firearms maker Herr Joseph Schulhoff (1824–1890), who gave up his simple life as a farmer and moved to Vienna at the age of forty-six in search of a change of career. His new repeating weapon contained a receptacle for holding ten cartridges. These could be poured directly and easily into the receptacle from a cardboard case. In the repeating action, working a knob handle fired the bullets, and to shoot a single bullet, a trigger was pressed.

The Maxim Gun (1887)

Meanwhile, in America, the famous inventor Sir Hiram Stevens Maxim (1840–1916) designed a very ingenious gun, which had the remarkable ability of automatically loading and firing itself by means of the force produced from the gun's recoil. The cartridges were arranged side by side in a belt which was fed into the gun by the recoil of the barrel. The gun also came with a metal shield for the protection of the operator.

The British Government ordered a sample, which was reported to be capable of firing 400 rounds a minute and 1,000 rounds in

four minutes, but instead the inventor submitted an improved model, as shown, which fired 1,000 rounds in one minute thirty seconds, and 2,115 rounds in three minutes forty-five seconds. The mounting of the gun was designed especially for the needs of Sir Henry Morton Stanley (1841–1904), the African explorer, and its rate of firing ranged from 660 to 670 shots a minute.

The Electric Lifeboat (1892)

The industrious Mr Edison, meanwhile, was busy developing a 'fabulous electric super-weapon' that could be sent out 2 miles ahead of a warship's bow, ready to blow up anything within reach. By 1892 this Sims-Edison torpedo had been launched. It was propelled through the water by the power of electricity, which was supplied to the motive mechanism by waterproof wires, fed out from a reel on the ship as the torpedo advanced.

Following the successful unveiling of his weapon, Edison resolved to apply the same principle to a contrasting application, and invented an electric lifeboat. His new vessel could be sent out for miles across the waves by means of an on-board electric motor, supplied with current by the waterproof wires that were fed out from the main ship behind it. Only two men were required to steer the electric craft, which was said to be as buoyant as the traditional oar-powered variant. Furthermore, once the lifeboat had reached its target, it could easily be reeled back to the mother vessel by means of the attached cables, thereby eliminating any risk of becoming lost in a stormy sea.

The Hydrophone Torpedo Tell-Tale (1893)

An enemy torpedo boat was the insidious foe that held ports and fleets across the globe in terror, but in 1893 a former mercenary Confederate officer, Captain Charles Ambrose McEvoy (*c.* 1828–1905), invented an automatic alarm that offered fair warning of their approach. This 'watchdog of the sea', as it was described, was named by its inventor the 'hydrophone', and was actuated by the vibrations of the waves caused by the screw of the enemy torpedo boat, or even from the torpedo itself. Trials carried out in the Solent by the government authorities found the invention sensitive enough to detect the passage of a vessel up to a mile away.

Captain McEvoy's apparatus consisted of two parts, one submerged at the point of outlook, the other stationed on shore or on board a ship. An electrical circuit, L L, connected these parts after the manner shown in the diagram. The submerged part consisted of an iron case shaped like a bell, which was plunged into the ocean, mouth downwards, to a depth of 40 or 50 feet, where it was anchored. The shape of the contraption meant that the upper part of the inside of the bell was occupied by air when submerged. Within this air space was fixed a copper box holding a delicate electric contact-maker, consisting of a platinum needle, N, resting its lower end on a platinum stud attached to the upper surface of a brass piece, P, fixed to the end of a horizontal spring, S. The vibration of the surrounding water, set up by the propeller of the approaching vessel, caused the needle to dance on the stud and rapidly interrupt the electric circuit. It followed that the current from the battery, B_1, flowing through the line, L L, and the shore indicator, became intermittent.

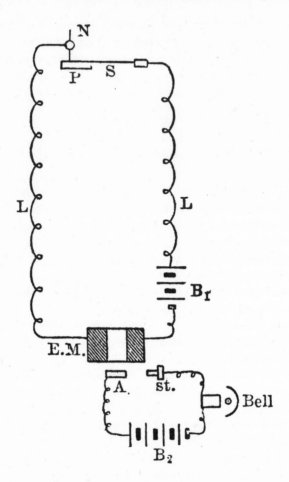

The shore indicator, which was called the 'kinesiscope', consisted of an electromagnet, E M, with an oscillating armature, A, and when the intermittent current passed through it, the armature swung to and fro until it came into contact with a magnetised stop, *st*, which arrested and held it. At this moment of contact, the circuit of a local battery, B_2, was closed, and an electric bell was rung, or an electric lamp was lit, to summon the attention of the watch. Thus the propeller of the enemy torpedo boat unintentionally announced its presence to the neighbouring vicinity.

The hydrophone was also employed to exchange messages between steamers at sea, provided the propeller of the sending vessel was stopped and started according to the Morse telegraphic code.

The Miniature Sun (1884)

Natural dangers, such as hidden rocks, were a different kind of hazard to seafaring vessels, and few waterways in the New World were more dangerous than New York's notorious Hell Gate. The passage through this narrow stretch of water, which connected Long Island Sound with the East River, was difficult and dangerous to cross owing to whirlpools and jagged rocks. By the end of the century more than 100 merchant and naval ships had sunk in this treacherous channel.

Once again it was electricity that offered a solution. In order to aid safer passage through the tidal strait at night, a gigantic artificial light was installed to guide sailors to safety. An iron mast 250 feet high was erected, and fitted on the top with a powerful electric light, so bright it became known locally as the 'miniature sun'. It illuminated not only the entire passage, but also miles of the approach on either side.

The Submarine Electric Lamp (1884)

In Greenock, Scotland, an interesting experiment was made on the *Tilly*, a small steamer built on the River Clyde for the Batavian fisheries. The ship was fitted with a powerful waterproof electric arc lamp of 15,000 candlepower, which could be lowered deep into the sea without any danger of electrocution, and was designed to attract the fish into the nets. Flexible conductors conveyed the current to the lamp, which could be raised or lowered by pulleys from the ship's side. The lamp was enclosed in a flint-glass cylinder 9½ inches in diameter by 14½ inches long. At the trial on the Clyde, the lighted lamp was submerged for four hours, during which time a bounteous supply of fish was caught, and the magical scene of illuminated water was said to be most beautiful.

The Illuminated Train (1885)

On land, the installation of powerful electric arc lights upon the front of the engines of railway trains, to be used as headlights during the night, was slowly becoming commonplace in America. By 1885 it was proposed to utilise these lamps in another way – namely, as backlights, so arranged that a flood of light could be cast backwards over and alongside the carriages just as passengers alighted upon arrival at a darkened station, when no other public lights were provided. It was believed that such an improvement would prevent passengers from missing their foothold and slipping through the gap between the train and the platform.

Back in Britain, trains on the Liverpool and Manchester line of the London & North Western Railway Company were permanently lit for the first time in around 1885. This was a huge undertaking for the authorities, as the locomotive had to be constructed so it not only pulled the train, but also worked a dynamo that generated the required current to power the incandescent lamps that were fitted inside the carriages. Double lamps were installed in each compartment, arranged so that if one were to break, the other would be automatically lit in its place.

The Electric Lamp Clock (1885)

Meanwhile, at the other side of the world, it was announced amid great excitement that the grand town hall of Melbourne, Australia, was to be fitted with a new electric lamp clock with large iron dials, of 13 feet in diameter. The hours were to be marked round the dial by incandescent lamps set like gemstones into the cup-like hollows in the rim of the face. At night these lamps were to be lit by electricity, and a row of smaller incandescent lamps were also to be set along each hand of the clock, so that a person at a distance would be able to tell the time during the hours of darkness.

The Asylum Gong (1887)

In the village of Virginia Water, Surrey, a new and somewhat curious electric installation was erected atop the sprawling Holloway Sanatorium, an imposing gothic-style building that was used as a Victorian asylum for the treatment of 'the insane of the middle classes'. It was built between 1873 and 1885, and in 1887 an electric gong, the largest ever to be built in Britain, was fitted at the top of one of the great lantern towers. Thankfully for the operator, he was not required to brandish the largest hammer in Britain; instead he simply turned a key, which switched on an electric current that automatically released the hammer. The boom of the gong was said to be audible for more than half a mile around. As the sanatorium was built within 22 acres of land, such a powerful gong was perhaps required in order to communicate at specific times of the day with the inmates, of which the number had risen to 600 before the end of the century.

The Electric Underground Parcel Exchange (1892)

The advancement of electrical engineering meant it was finally possible for late Victorian inventors to turn even their most outlandish visions into reality. In consequence, all sorts of new and curious public conveniences were introduced which offered imaginative solutions to contemporary problems.

Traffic congestion was one such predicament which had become particularly bad on the streets of London and other large cities. In an attempt to avoid blocking the thoroughfares it had been necessary to prohibit the collection and delivery of parcels in busy districts during business hours. By 1892 these restrictions were beginning to have a negative impact on local companies, so in an effort to surmount these complications, the electrical

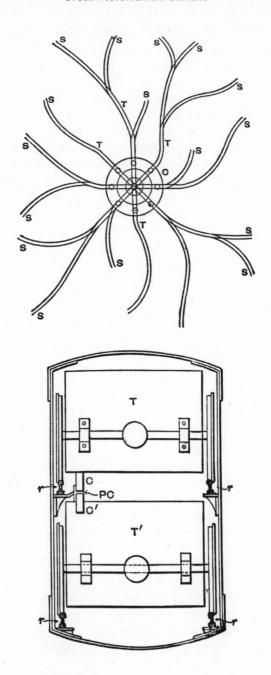

engineer Mr Alfred Rosling Bennett M.I.E.E. (1850–1928) developed a new system of conveying parcels from building to building via a miniature underground electric railway. He envisioned that if such a system were permanently introduced underneath the larger cities, the vehicular traffic of the streets above would be greatly diminished.

The top image, opposite, represents a general plan of his lines, which were small tubular railways, laid many feet below the ground in replication of a telephone exchange. The central exchange of the railway, which was staffed by operators, is shown at C, to and from which the parcels were circulated by the branch lines, T T, to the homes or business premises of the subscribers to this service, S S.

The bottom image, opposite, shows a cross section of the tubes in which T, T^1 were the parcel trains, running on rails, *r r r r*, by means of electricity conveyed along parallel conductors, P C. The electric charge was collected from these by brushes, C, C^1, to feed the electric motors driving the trains.

The Electric Parcel Post (1884)

The authorities at the Royal Aquarium and Winter Garden at Westminster tested an overland variation of the electric railway, which had been adapted for conveying letters, parcels, and other lightweight goods from one depot to another. It was the design of a Danish engineer named Mr Frederick Ludewig Hahn Danchell. The novelty of his system was the unusual positioning of the rails, which ran one directly above the other, the carriages travelling in perfect balance between the two. The wheels of the carriage, fitted in a single-file line and powered by an on-board electric motor, were deeply grooved in order to clutch the rails on either side, preventing the car from leaving the track, no matter how sharp the curve or high the speed. Mr Danchell expected that the cars could run on his railway at velocities of between 150 and 200 miles per hour, thereby providing a speedy system that could deliver parcels across the whole of the country.

The Electro-Photographic Thief Detector (1892)

It was, perhaps, a humble American cigar merchant who demon-strated one of the century's most inventive uses of electricity after he repeatedly missed expensive cigars from a show cabinet in his office at Toledo, Ohio. Having employed detectives to watch the premises in vain, he at last hit on the plan of automatically photographing the thieves by way of an instantaneous camera and flashlight. This forward-thinking gentleman dubbed his innovative system the 'electro-photographic thief detector'. An electric circuit was so arranged that when the cigar cabinet was opened, the thieves would close the circuit and send an electric current to a box which contained a hidden camera directed on the case. The current, by means of electromagnets, not only lifted the shutter of the lens, thus exposing the camera for a short time, but also struck a match on a rough disc to ignite the flashlight.

The ingenious device was installed in the merchant's shop in 1892, exactly fifty years before the German engineering company Siemens AG was credited with introducing the world's first closed-circuit television system.

One night, shortly after the electro-photographic thief detector was put in place, two boys stealthily entered the cigar merchant's office. They opened the case to steal some of the goods and were promptly photographed. The picture led to their apprehension, trial, and punishment.

The Electric Burglar Alarm (1884)

Over in France, an attempt was made to produce an artistic design for a domestic night watchman in the form of a *dragon vigilante*. The opening of a door or window in the vicinity of this bronze statuette actuated an electromagnetic arrangement concealed inside the dragon, which caused a clapper to hit an alarm bell, thereby waking the house.

The Electrical Trumpet (1887)

Another kind of domestic bell was the electric doorbell, which, by the end of the century, had become commonplace across Europe. However, the harsh, raucous sound it made was objected to by many residents, prompting a French inventor named Dr Legang to replace the ordinary doorbell with an electric trumpet. It consisted, as shown in the figure, of an electromagnet contained inside a brass tube, T, just over 24 inches long by 1½ inches in diameter. A soft iron armature, A, was mounted opposite one pole of the magnet on a metal diaphragm, D, which, on vibrating, made and broke contact with the screws, S. The screws, the diaphragm, and magnet, were all connected to a battery, so that when the current was turned on, the diaphragm, D, vibrated and emitted a musical trumpet-like sound, which some people found to be more pleasing than the shrill ringing of a bell.

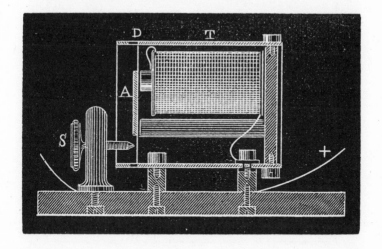

Electrical Utensils (1892)

At the Crystal Palace Electrical Exhibition of 1891 there were numerous examples of how traditional domestic utensils could, by that time, be heated by electricity. This was based on the

principle that when an electric current was sent through a high resistance – for example, a fine wire of platinum – a copious amount of heat was generated, and very quickly.

The electrical engineer Herr Gustav Binswanger (1855–1910) ingeniously applied this scientific fact to ordinary kettles, frying pans, flat-irons, and so on, by embedding the platinum wires in a special cement which insulated them from each other and from the metal of the utensil. A kettle was thus rendered electrical by simply coating the bottom with the cement, embedding the platinum wires in it, and connecting their ends to binding screws. When the householder connected these screws to the leads of a domestic electric light installation by flexible conductors, the current flowed through the platinum wires, and conveniently heated the kettle without the use of a flame.

Another popular attraction at the exhibition was an electric frying pan, which could cook an omelette in two minutes at an estimated cost of one-twentieth of a penny for energy consumption. The heat was produced in the same way, and demonstrated that by using this modified utensil, no soot or smoke would spoil the dish, as there was no need to light a fire.

The Electric Chicken-Hatcher (1884 & 1893)

Many Victorians had access to their own supply of fresh eggs, and for the benefit of farming families, who kept a considerable

number of chickens, a French engineer devised an artificial egg-hatcher, which was heated and controlled by electricity. The heat was produced using a series of bare platinum wires made incandescent by the electric current. A regulating thermometer kept the eggs inside the electronic hatcher at a constant temperature, for it ensured that the current could be automatically cut off when the temperature became too high. Once the mercury column within the thermometer rose to a given height, this would short-circuit the poles of the battery that supplied the current.

A later variant of the electric chicken-hatcher is illustrated on the previous page. As can be seen, there was a large chamber with a glass lid above, in which the cold eggs were dried and warmed before being carefully placed in the partitions of the drawers underneath for incubation. These partitions were of different dimensions, allowing the eggs to be classed according to size. The hot air from a heating lamp travelled through a pipe or flue on the right-hand side of the incubator, an arrangement that offered a large heating surface.

The Electrical Wedding (1892)

By 1892 the novelty of electricity had inspired one thoroughly modern American couple, ready to embrace the technological delights of the new century, to get married at an entirely 'electrical wedding'. Electricity had been installed at the Shrine of Hymen, where their nuptial celebrations were enjoyed. As soon as their guests sat down, clusters of multi-coloured glow lamps that were hidden among the foliage brilliantly lighted up the room. The ringing of electric bells and the playing of electrical instruments hailed the advent of the bride and groom.

After the first course of the wedding breakfast, the lights in the room were dimmed and miniature glow-lamps, concealed among the flowers on the table, began to twinkle like glow-worms and fireflies. Translucent vases of the finest glass and porcelain were also lighted up from within, and a tiny lamp could be seen glowing in the bride's hair. A toast having been given, two electrically powered serpents uncoiled from one of the vessels on the table, and confronted the new couple; but what this movement was intended to signify must be left to conjecture.

Coffee prepared by an electrical heater was served at the table after the meal; the speeches were loudly applauded by an automatic claque in the shape of an electrical kettledrum placed under the table, and as the company dispersed, the electric current fired off a pyrotechnic display in the garden.

'It is perhaps needless to add that the wedding took place in Baltimore,' said a reporter who attended the event.

Voting by Electricity (1884)

History records that it was the Americans who were generally quicker to embrace electrical technology than the British. Testament to this was the fact that in 1884 an electric vote-recording machine was introduced in Washington, while in Britain the traditional systems of pencil and paper, or a show of hands, remain steadfast to the present day. Despite some initial concern, the electorates were reassured that all danger of error had been thoroughly guarded against, and furthermore the voter did not even need to leave his seat to cast his vote. At the desk of each elector was a small lever, which when turned to the right registered 'Yes', and when to the left, 'No'. The vote was automatically recorded on a piece of paper in the tellers' room, and a bell and annunciator further signalled the decision.

Over the years there have been numerous attempts to introduce national electronic voting systems, but despite the inventor's insistence of its infallibility, there have been an array of technical errors associated with such arrangements. During the presidential election of 2012, the system of automated polling went infamously wrong when the touch-screen machines in Pennsylvania kept erroneously registering a vote for Mr Mitt Romney whenever the voter pressed the name of any other candidate. This serves to show that even the most cutting-edge technology is not always a match for the unfaltering pencil.

6

LIGHTS, CAMERA, ACTION!

The Electric Light Spectacles (1884)

The introduction of the electric light had triggered much discussion regarding the potential danger to eyesight posed by these dazzling new lamps. Artificial illumination was such a novelty to the Victorians that they found their gaze being constantly drawn to the bright lamps, even though doctors advised that staring into the light was not conducive to the health of the eye unless the bulbs were shaded by ground or opal glass globes. Physicians and electrical engineers were in agreement that one day, when the electric light had ceased to be a novelty, people would not stare at it any more than they stared at the sun. Until such a time, it was hoped that the tinted 'electric light spectacles' introduced by Dr William Henry Stone F.R.S. (1830–1891) would provide sufficient protection.

These special spectacles had blue glass in front and red or 'blinker' glass at the sides, which could be folded across over the blue ones. There were several different kinds of bulbs in use in the late Victorian era, as inventors vied to perfect the technology. The incandescent lamp, championed by the likes of Sir Joseph Wilson Swan F.R.S. (1828–1914) and Mr Thomas Edison, was to be looked at through the blue glass, and the carbon arc lamp of Sir Humphry Davy (1778–1829) through the blue and red together. This was because it was assumed that the red rays emitted from the incandescent lamp were likely to cause injury to the eye, and that the blue or violet rays were to blame in the case of the brighter arc lamp.

The lenses used in these safety spectacles were selected and combined by aid of the spectroscope, a scientific instrument used to measure the colours of the spectrum, and thus the exact tints were obtained. These particular colours were believed to temper and cut off the dangerous rays in the two cases, rendering it safe to stare straight at any electric bulb.

Electrical Light Effects (1884 & 1885)

By the late nineteenth century lavish exhibitions had become a regular occurrence in Victorian England, providing a platform for hundreds of national and international exhibitors. Families flocked to these events, eager to feast their eyes upon the latest cutting-edge technology that the world had to offer. London's International Health Exhibition of 1884 boasted some astounding effects of illumination, which were admired each evening. An enormous fountain was lighted by incandescent lamps in such a manner that the water glowed, and as it fell the spray seemed to consist of golden drops. To enhance the spectacle the lights were projected upon the waters through panes of coloured glass.

The following year, the Inventions Exhibition was held at London's Albert Hall galleries. In addition to the illumination of the fountains by electric arc lamps, which created such a conspicuous effect at the Health Exhibition, an electrical engineer named Sir Francis Bolton (1831–1887) arranged for the illumination of the entire grounds of the Albert Hall and the surrounding buildings by means of small 5- and 10-candlepower incandescent lamps in coloured glass bulbs – red, green, blue and yellow. These little lights twinkled like fairies among the leaves of the foliage of the shrubs and trees, amazing everyone who set their eyes upon them. There were no fewer than 9,020 bulbs running on insulated wires throughout the grounds, taking the place of the small oil and tallow lamps that were previously employed at fêtes for this type of illumination. Being electric, these new lights could be mounted on the twigs of even the highest trees, and did not require daily renewal. Some of them were also arranged in the ponds to illuminate the flowers of aquatic plants.

The Incandescent Streetlamp (1892)

The traditional arc lamp, though a once splendid street light in a clear atmosphere, such as that of Paris, was not so visible in fogs and mists, owing to the absorption of its light by the water vapour in the air. This kind of illumination had been in

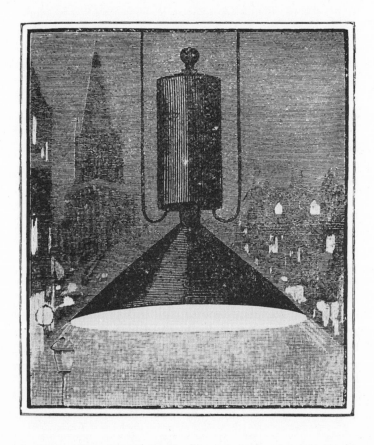

existence since the turn of the nineteenth century, when the English chemist Sir Humphry Davy invented his aforementioned arc lamp by wiring two carbon rods to a battery, which began to glow a brilliant white light.

The new incandescent electric lamp, however, which was invented in the late 1870s by Edison, was found to be not so sensitive to fogs, hence the 2,000-volt street lamp, which was introduced by the electrical engineers Messrs Swinburne & Co., eventually proved very useful for the towns and cities of misty Britain.

This innovative street lamp is shown in the figure opposite, where a small transformer was placed over the lamp and its enamelled iron shade. The volts passed into the transformer by the wires shown, and the transformed current flowed to the filament of the lamp, heating it white-hot and yielding a light of 32-candlepower.

The Hinged Lamp Post (1887)

Lamplighters were once a common sight as they walked the cobbled streets at twilight to kindle the old-fashioned gas lamps. However, their services were no longer required once electric street lights were installed, and hence these new lamps could be raised much higher above street level. Though the incandescent street lights received general approval from the public, their height often rendered them difficult to clean and repair. This led to the development of electric lamp posts that were hinged in the middle, therefore no ladder was required to reach them.

In some older parts of town, these new hinged lamp posts were also adopted for gas lamps, as they could be easily lit by bending them to a horizontal position – the lamplighter carrying a key for the purpose – and then returned upright again.

The Crime-Fighting Streetlamp (1887)

The accompanying illustrations represent a very innovative lamp post, which was introduced to the streets of New Haven in

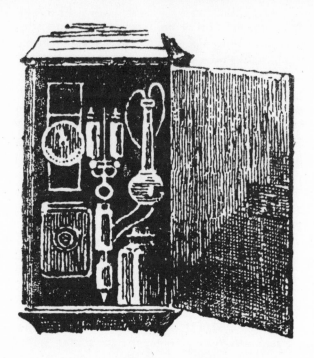

Connecticut, United States, by Messrs Brewer and W. C. Smith in a desperate attempt to restrain the growing criminal classes of the district. The secret to its novel crime-fighting approach could be found inside the body of the lamp post in the form of a hidden telephonic apparatus, as shown above. When important information was to be relayed to a patrolling policeman, his colleagues at the station could telephone the lamp post nearest to him and speak to the officer directly. If a crime had been committed in a particular part of the town, all the policemen in the neighbourhood could be informed of it via the local lamp posts, and thereby co-operate in the capture of the felon. Additionally, a simple device for raising and lowering a red screen of glass around the flame of the lamp was fitted to each post, thereby emitting a bright red signal light. This was operated by electricity at the local police station, and was used to quickly alert all the local policemen to an emergency.

The opposite image shows the crime-fighting streetlamp in use.

Despite the best efforts of Messrs Brewer and Smith, crime rates continued to rise in New Haven, and by the 1990s crime levels had risen to such an extent that the city ranked sixth in violent crime per capita in the whole of the United States.

The Signal Light of the House of Commons (1893)

Back in London, a new signal light on the clock tower of the House of Commons was introduced in order to alert Members

of Parliament who regularly frequented the West End clubs that the House was sitting. It was installed in a camber above the famous bells, 250 feet above street level. The lantern, shown in the figure above, consisted of highly polished refractive lenses surrounding a lighthouse burner of sixty-eight gas jets, and produced a mass of flame 8 inches in diameter and 6 inches high, which had an illuminating power of 2,400 candles. A glazed lantern, 12 feet high and 9 feet in diameter, enclosed the whole.

The old lamp that it replaced could only be seen through the arc of a circle of 210°, corresponding to the West End district, and it was perhaps a sign of the times that this new light was visible all around, which was found particularly beneficial to those Members who had ventured further afield.

Darkness Photographs (1887)

Another indication of the changing times was the sudden availability of a wide range of artificial lighting for city dwellings. Gone were the days of shadowy candlelight, for now one's home could be illuminated by gas, oil, or electric lamps, of all shapes, sizes, and designs. The progress made by engineers in artificial illumination cleared the way for inventors to experiment with a relatively modern form of technology that relied entirely on light. This was, of course, photography.

The camera obscura, more commonly known as the pinhole camera, had been a form of entertainment since ancient times, but it was not until the nineteenth century that the art of photography truly began to evolve. The very first cameras to actually take photographs, rather than just project images on to a screen, were invented in the 1820s, but the world had to wait until 1884 for the first reel of photographic film to be devised.

From the 1870s until the early twentieth century many cameras were fitted with gelatine plates, upon which the light from the image was imprinted, and in 1887 Captain William De Wiveleslie Abney C.B., R.E., F.R.S., etc. (1843–1920), a chemist by trade, stunned photography enthusiasts by claiming to have prepared photographic plates so sensitive that they were responsive to the invisible heat rays beyond the red end of the spectrum. In other words, instead of relying upon light, Captain Abney's futuristic camera was capable of capturing the image of a heated body, even in the dark, thereby producing what were dubbed 'darkness photographs'.

The captain was clearly a man well ahead of his time, as the history books reveal that the first thermal imaging camera was not invented until 1929, when a Hungarian physicist named

Kálmán Tihanyi (1897–1947) developed a special camera for use in Britain's anti-aircraft defence.

Hospital Photography (1884)

As photography developed, its uses multiplied. By the 1880s, for instance, certain hospitals in France had erected adjoining photographic studios for the purpose of photographing patients at different stages of their treatment in order to better study their conditions. A simplistic method of taking photographs known as the rapid dry-plate process was employed, and Professor Jean-Martin Charcot (1825–1893), the founder of modern neurology who worked and taught at the Salpêtrière Hospital, devised an electrically operated camera for taking a series of views in rapid succession. Patients were photographed on entering the hospital, and at regular intervals afterwards, such, for example, as hysterical patients whose conditions required close observation. The photographs were duplicated for the general use of medical professionals in studying the diseases and disorders, and copies were also kept in albums within the hospital for future reference.

Professor Charcot spent much of his time diligently studying his images of patients who were suffering from various neurological diseases, such as Parkinson's disease. Through his dedication many nervous disorders, including multiple sclerosis, were identified. A series of Charcot's medical photographs was published between 1876 and 1880, entitled *Iconographie Photographique de la Salpêtrière.*

Photographs of Lightning (1885 & 1887)

Some of the earliest and clearest photographs to be taken of a bolt of lightning were captured by a German photographer named Dr Kayser, who exhibited his images in Berlin before the Society for the Furtherance of Photography. One of his pictures, taken on 16 July 1884, depicted a clear flash of lightning with an

estimated thickness of 4 feet across the main discharge. When magnified four times, it became apparent that the lightning bolt consisted of four distinct parallel currents, with alternating dark and bright stripes between the first and second currents. This striated appearance had never been clearly shown before, and Dr Kayser hoped that his photograph would prove instrumental in the study of this electrical phenomenon.

A report published in the *Aberdeen Journal* on Tuesday 28 October 1884, describing the doctor's groundbreaking picture, went on to reflect, 'It is to be hoped that the lightning will not take it into its head to object to being thus critically examined, for if the dangerous element "turned nasty" it might make matters particularly unpleasant the next time Dr Kayser and his scientific friends happen to be out in a thunderstorm.'

The accompanying figure is a reproduction of a very fine photograph of lightning, which was reportedly taken by Monsieur Schleusner at 10.30 p.m., pending a terrible storm that struck at the Château Rougemont in Tours, France, on 26 May 1887. Photographs such as this served to prove for the very first time in history that the conventional representation of lightning, namely a zigzag line, was a misrepresentation.

The Tornado Photographed (1885)

The following image represents one of the earliest known photographs of a tornado by a photographer named F. N. Robinson. It was taken in the afternoon on 28 August 1884 at Howard in the Miner County, South Dakota. The tornado was said to have passed over the town and destroyed everything in its course, killing several people.

The lower line of the photograph represents the prairie over which the storm passed; the upper dark mass was the cloud-belt which accompanied it. The resemblance of the storm to a waterspout was striking to Victorian viewers, who prior to this date had never been fortunate enough to see the formation of a tornado in such detail.

Deep-Sea Photography (1887)

A gentleman named Mr William Thompson was credited with capturing the first ever underwater photograph in 1856. At that time, however, the technology was not particularly advanced and the photograph was said to be rather unclear.

By 1887 submarine experiments were underway once more, when the French attempted to take well-defined underwater photographs for the first time. Their arsenal consisted simply of an electric incandescent light and an ordinary camera sealed inside a watertight box. Sunken ships were the proposed subjects, and it was expected that, if the technology could be developed, such cameras would prove useful to deep-sea divers of the future.

Sadly for the French, these marine experiments did not quite go according to plan, and the advent of underwater photography did not come about until 1893, when the Versailles-born photographer Monsieur Louis Boutan (1859–1934) developed fully waterproof cameras complete with flashlights.

Long-Distance Photography (1887)

It was the French who were leading the way in the race to perfect the art of photography. Before the 1890s a photographer named Monsieur Marey discovered that it was possible to capture instantaneous photographs in a 2,000th of a second, and he hoped to reduce this still further.

Another Frenchman, Monsieur Lacombe, succeeded in taking the first long-distance photographs by the simple fixture of a telescope in front of the lens of his camera. This was held in position by means of a screw cap, shown at point A below, which

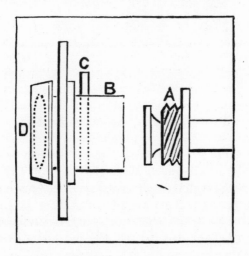

screwed into the mounting of the lens at B. A diaphragm with a large aperture, placed at C, kept it from abutting on the lens at D. With this device, Lacombe was able to take photographs of the windows of distant houses that were up to a mile away, and also capture faraway views of picturesque scenery. His new discovery was expected to be a tremendous tool for Victorian tourists who were keen to encapsulate the distant landscape.

Also popular with travellers was a very small instant camera, which was brought out in Paris. It weighed about 700 grams, and, as illustrated above and opposite, resembled a flat, round box. It was simply held in the hand and operated – the likeness being

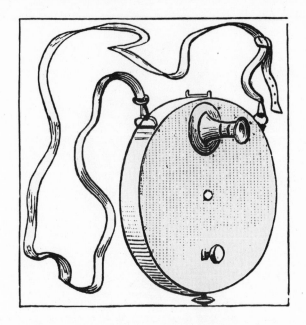

taken instantaneously. Of added convenience was the fact that the camera could hold several photographic plates at any given time, meaning that the equipment did not have to be replenished after every click.

The Opera-Glass Camera (1893)

By 1893 another French inventor by the name of Monsieur Jules Carpentier (1851–1921) had devised an 'opera-glass camera', so called because of its resemblance to a traditional pair of opera glasses. It contained twelve small photographic plates, and was operated by holding the device to the eyes and pointing it towards the object to be photographed. The image on the ground-glass plate was seen through the aperture, C (as shown overleaf), which was covered by red glass. The plates were held in a drawer, A, and when the photographer was ready to expose one to the light, it was pushed into place by a finger piece, B. An enlarging apparatus also formed part of the camera.

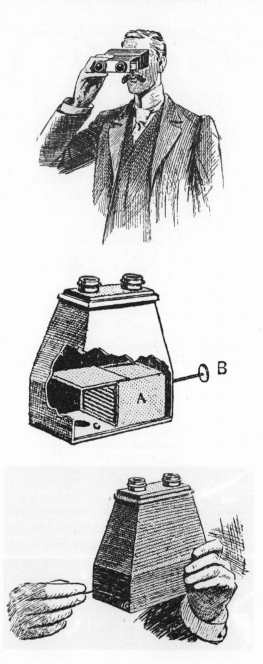

The Camera Obscura in Coastal Defence (1885)

Such modern photographic gadgets proved a great novelty with the Victorian public, who relished the opportunity of having their own and their loved ones' likenesses captured in a flash, to keep for posterity. In previous decades such a luxury would have only been available to those who had the time, money, and inclination to commission an artist, who would have had to spend a considerable period of time painstakingly painting or sketching their subject without any guarantee that the finished piece would turn out entirely true to life. During the Victorian era photographic studios were set up across the country, and families would flock in their thousands to sit and have instantaneous portraits taken.

While photography had been brought within the reach of the common man via commercial high-street studios, military intelligence suggested that forces shadier than a photographer's darkroom were at work within the wider world, and photographic technology, based on its fundamental principles, was ready to fight back.

France had been a perceived threat to Britain's national security for many centuries, and relations were particularly strained by the end of the Victorian age. With the conflict against archenemy

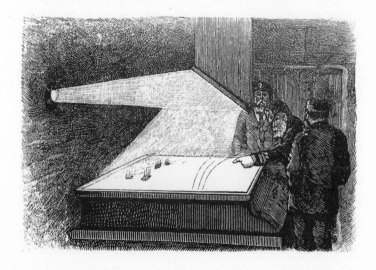

Napoleon Bonaparte (1769–1821) still in living memory, and
Prussia exposing itself as a militaristic and ambitious kingdom
following its triumphant victory in the Franco–Prussian War of
1870–1871, Britain found itself on high alert. In 1882 matters
were made worse by the planned construction of a great railway
tunnel underneath the English Channel, permanently linking
Dover with Calais. This pronouncement, which was intended
to improve trade links between Britain and the Continent, only
served to cause further alarm among a mistrustful public, who
feared the construction of such a passageway would be like
opening the floodgates to invaders from France, Germany, and
any other nation who suddenly took a fancy to this scepter'd
isle.

Incensed by this dreadful possibility, Victorian inventors took
to their drawing boards and came up with a cunning solution
that would help to protect British harbours in the highly likely
event of an enemy invasion by sea. The key to this plan was the
simple arrangement of the camera obscura, which was utilised
to guide a firing party in the manner shown in the engraving.
The officers' observation table was made from a sheet of glass, on
which the pinhole camera threw an image of the neighbouring
port or harbour by means of the optical arrangement shown.
Black points on the image indicated the positions of British
torpedoes. The enemy's ships – or, rather, the images of them –
would be seen outside these torpedo lines. When a ship appeared
to approach a torpedo, the officer on the lookout would press a
corresponding key, sending an electric current to explode that
particular torpedo, hence thwarting the invasion and keeping the
shores of Britain safe for a little longer.

The Heliochromoscope (1892)

In their quest to capture lifelike colours, photographic engineers
soldiered on with their experiments, and by the end of the
century they were set to overcome the limitations imposed by
monochrome images. Since the 1840s inventors had been busy
working on a reliable process for achieving this naturalistic

effect, and over the decades many offered their own, often flawed, theories as to how this could be accomplished.

Long before the introduction of colour cameras, a common method used to apply shade to black-and-white prints was quite simply to paint over them by hand. In around 1865, however, Mr Henry Collen (1797–1879), the painting master to Queen Victoria, suggested a method for capturing the natural shades of an object without the aid of the brush. He hypothesised that colour photographs might be obtained by the composite process of making three pictures: one for each of the three primary colours, and cementing one over the other on a white background. Though this was not an original idea, and many attempts to realise this method had ended in comparative failure, by 1892 an inventor from Philadelphia named Mr Frederick Eugene Ives (1856–1937) had claimed victory.

According to the old theory of light advocated by the eminent physicist Sir David Brewster (1781–1868), the primary colours were red, yellow, and blue, and a mixture of these was believed to produce white light. However, the German physician, Herr Hermann von Helmholtz (1821–1894), was able to prove that instead of primary colours, there were primary – or fundamental – colour sensations, due to the presence of three kinds of visual nerves in the eye. These simple sensations, Herr Helmholtz asserted, were red, green and blue. All other colours were compound sensations; yellow, for example, could be produced by a mixture of red and green, and the yellow rays of the spectrum appeared so only because they had the power to 'excite' both of these fundamental sensations at once, and equally.

It therefore followed that if it was possible to take three photographs of an object – one reproducing the effect upon the nerves of the red sensation, one the effect upon the green, and a third the effect upon the blue – these might be combined into a single picture that would reproduce the original tints of the object. Hence it was necessary to take transparent images and to illuminate them by coloured light. This was done by means of translucent coloured screens, and it was here that Mr Ives made one of his great advances on his predecessors in the same field. His screen of yellow glass, for example, interposed between the camera and the object, allowed the yellow rays to pass through to the camera and cut off the others. It was the yellow rays, therefore, that reached the sensitive plate and produced the impression. Three transparent pictures of the same object were thus obtained through three different-coloured panes of glass, and by a new 'triple camera', invented by Mr Ives, they were taken on the same sensitive plate by a single operation. This remarkable new invention became known as the 'heliochromoscope'.

At the time, it was thought probable that a means would eventually be worked out for printing the photographs in the natural colours of the objects, but until such a process was perfected, the pictures taken by this new camera could be projected on to a screen by means of a magic lantern. In a lecture at the Royal Institution, London, Mr Ives projected a number of lifelike views taken by himself in the Yellowstone Park in the American state of Wyoming. He also exhibited a small handheld apparatus for reproducing the pictures. This was fitted with an eyepiece and was about as large as a drawing room stereoscope – an optical device which, when looked through, gave the illusion of a three-dimensional image. On looking into the eyepiece the observer saw a 3D picture, such as a bouquet of flowers, reproduced in its natural tints.

Coloured Photographs on Glass (1884)

By 1884 it was said to be possible to produce delicate coloured photographs on glass by fixing a paper photograph on to the back of a cushion-shaped piece of glass using a type of transparent cement. When this had dried, two-thirds of the thickness of the photograph was rubbed away with sandpaper until the photograph was reduced to a fine film. This film was rendered transparent by soaking it in melted paraffin wax, and then applying translucent coloured paints, which appeared soft and subtle when seen from the front, on to the paraffin layer. The background, and heavier portions of the photograph, were then painted on the front of another cushioned-shaped piece of glass, and fixed behind the first.

Through trial and error, a lady named Mrs Nelson Decker found that this second piece of glass could actually be dispensed with. Her improved method involved dipping the photograph in paraffin after the transparent colours had been applied, and the heavier colours could then be painted on the back of this second coat. A third layer of paraffin was then applied, and the background painted on to that. A final coat protected the whole against the deteriorating effects of the atmosphere.

Spark Photography (1893)

By the end of the century it was not just the differing photographic techniques that were being scrutinised, but also the
subject of the photograph. After much study it was discovered
that, by means of a quick shutter speed and a bright electric
spark, the shadow of any changing or rapidly moving object –
such as a breaking mug, a flying bullet, or a bursting soap bubble
– could be photographed with great precision, when hitherto
the image would have appeared blurred. This technique became
known as 'electric spark photography'. Such was the intensity of
the spark that all efforts to capture an image of the actual object,
rather than just the shadow, had failed, as the photographs were
overexposed. However, it was later found that this could be
achieved by reflecting the light of the spark on to the moving
object with the use of a revolving mirror and a very quick shutter
on the camera. It was claimed that, by this means, images of
objects travelling at rates of up to 180,000 miles an hour could
be clearly captured without any motion blur.

Photographing the Walk (1892)

By the close of the century the instantaneous photograph had
become a recognised aid in both science and art for the study of
animal, vegetable, and even mineral movement. The unassisted
eye was unable to follow the different phases of rapid change,
for example the track of a lightning flash, the beating of a
bird's wings, the swirling of leaves, or the splashing of water.
Photography had already proved to the Victorians that the
conventional zigzag flash of lightning was simply a figment of
the imagination, and such depictions were already disappearing
from works of art. It was also revealed through photography
that artists had hitherto been misled by the eye when drawing
galloping horses; the positions of the legs in certain works of
art were, in fact, impossible. This led certain contemporary art
historians to reason that when the spectator of a work of art
was sufficiently educated to detect a fault that was contrary to

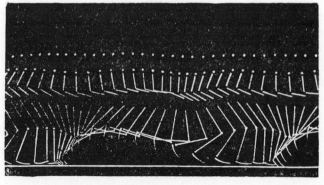

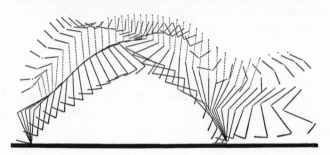

nature, his pleasure in the work was marred, and in some cases, destroyed completely.

'It is possible for an artist to become so steeped in science as to grovel and thus lose his artistic sense of the beautiful,' they analysed.

'But, on the other hand,' their detractors argued, 'a regard for science by correcting his imagination and judgement will keep him true to Nature, and prevent him from blemishing his work with errors.'

Around the year 1892, Monsieur Georges Demenÿ (1850–1917), an inventor and assistant to the eminent scientist Monsieur Étienne-Jules Marey (1830–1904), was busy making a complete study of the movements of a man's walk by means of tiny incandescent lamps attached to his attire, which was tight and made of black cloth, as shown in the top image on the previous page. By photographing these lamps as the man walked or ran against a dark background, the varying positions of the body and limbs were clearly observed when combined into a single image. The middle image on the previous page, for instance, shows the successive images obtained in this way from a runner; the white dots and lines indicating the positioning of his limbs.

The bottom image on the previous page was the analysis of a running high jump, obtained on a fixed plate by means of twenty-five images captured per second. Though some sceptics described the diagram as 'somewhat eccentric, not to say demoniac', it was possible for Monsieur Demenÿ to draw useful conclusions from it. It is clear from the right of the image that the skeletal gymnast was preparing to leap over the invisible hurdle, and from the manner in which his spidery limbs were drawn up in the middle of the picture, he was probably clearing the bar. The curve of descent is plainly visible on the left-hand side.

This interesting selection of photographs would have probably surprised a professional Victorian high jumper, who could not have been expected to visualise the figure he cut in the air, but thanks to the invention of this zoetrope-esque camera, which combined the several views in one, the whole picture was harmonised.

The Phonoscope (1892)

One of Monsieur Demenÿ's next challenges was to introduce a new way of teaching deaf-mute children to speak, and this, he hoped, could be achieved by means of a magic lantern. His objective was to take a series of rapid photographs of the faces of a teacher or lecturer as they spoke, and then combine the successive pictures using a zoetrope, which reproduced as an optical illusion all the motions of the mouth, and the changing expressions of the face, as the speaker spoke. As deaf-mutes were accustomed to reading what a person said by the movement of

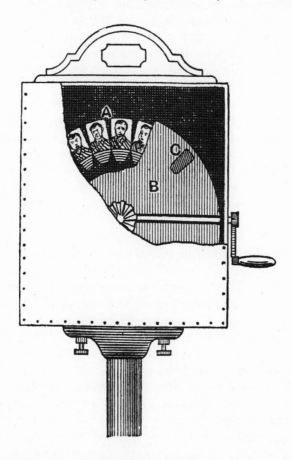

their lips, it was hoped that they would be able to do the same from studying these 'speaking photographs'.

The image on the previous page provides a view of Demenÿ's 'combining apparatus', or 'phonoscope', with part of the back removed in order to show the inner workings. The transparent positives were arranged in series around the periphery of a revolving disc, A, which was turned by a handle. At the same time, a circular screen, B, which had a single aperture or window, C, was also revolved at a much higher velocity. An eyepiece allowed the observer to see the photographs through the window, and the speed of the plates was so adjusted as to combine all the images into one, giving the appearance of animation.

Demenÿ began by capturing his own face speaking the phrases '*Je vous aime*' and '*Vive la France!*' Unknown to him, however, his sentences were not fully photographed and as a result his experiment failed due to the students not being able to follow what was being pronounced.

This did not dampen the public enthusiasm for such innovations, and speculation was rife about whether or not the technology of the future would offer 'talking photographs'. It was envisaged that one day people might even possess family albums containing pictures of people who would not only be able to smile, but would also be able to speak, or at least appear to speak, as if they were alive.

Moving Photographs (1887)

The ability to film a motion picture was the zenith of technology that occupied the thoughts of the architects of the day, and in 1887 the German photographer and inventor Herr Ottomar Anschütz (1846–1907) devised a new photographic apparatus for reproducing on a screen the moving image of an object. This invention, constructed especially for the Prussian war minister, was dubbed the 'électrotachyscope', otherwise the 'tachyscope', and was demonstrated in public for the first time at the Chicago World Fair of 1893.

In order to operate the apparatus, Herr Anschütz first took a series of rapid photographs of a moving object by means of highly sensitive plates, receptive to an exposure of around a 5,000th of a second. The camera he used for the task possessed twenty-four lenses, which meant that the same number of pictures of the moving object – say a horse leaping, or a marching band – could be taken successively in the space of around three-quarters of a second. As each of these photographs depicted a slightly different posture of the moving body, Anschütz reasoned that by causing the images to succeed one another in rapid succession on a screen, the projected image would appear to move, reproducing the exact same movement performed by the object. As the persistence of images on the retina would bridge the intervals between each photograph, the motion would be smooth on the eye.

The pictures Anschütz captured with his camera were very small, so they were enlarged after development and transferred on to thin glass plates. The twenty-four transparent views thus obtained were mounted on an iron disc containing twenty-four apertures near the rim. This disc was then rapidly revolved before an early type of neon light known as a Geissler vacuum tube, which sent a spark of light through the tube as each picture reached the line of projection, thus illuminating the glass. This caused a succession of bright images to be thrown on to the screen one after another. As the eye merged each individual image into one, an astounded audience witnessed the apparent movement of the projected object.

Despite the fact that Anschütz was developing his movie camera in 1887, the earliest known surviving motion-picture footage dates back to the following year, 1888. This rudimental film depicts the now world-famous Leeds Bridge, and was shot in the West Riding of Yorkshire by the photography pioneer Monsieur Louis Le Prince (1841 – d. unknown).

Monsieur Le Prince mysteriously went missing in France in 1890 after visiting his brother in Dijon. He was last seen boarding a train on his way back to England where he was to obtain a patent for his groundbreaking new camera ... but this was never to be, for Le Prince was neither seen nor heard from again. Though his luggage had seemingly vanished from the train nobody witnessed the filmmaker alighting, and he failed to arrive at his intended destination. His body was never discovered, and his unexplained disappearance remains a mystery to this day. Monsieur Le Prince's departure cleared the way for his American rival, Thomas Edison, to be credited with inventing the world's first truly operational movie camera a year later.

The Optical Performance (1893)

A contemporary of Le Prince named Monsieur Charles-Émile Reynaud (1844–1918) developed his own motion-picture technique based on the same principle as the zoetrope, by means of a large number of instantaneous images taken consecutively. Monsieur Reynaud began by taking sequential photographs of a theatrical performance by live actors. These images were then impressed on a long strip of photographic film, and their number was such that every successive pose of the actor was captured. When this film was run through a zoetrope, and the images projected on to the screen one after another so quickly that the eye could not detect any interval between them, the

images blended into one, and gave the illusion of animation to the figures. Thus it was discovered that all the action of the traditional theatrical production could be represented by the magic lantern after the manner shown in the illustration, where the operator was employed to work the apparatus behind the screen.

Following the triumph of his performance, Reynaud transposed the same principle to drawings, and was ultimately responsible for showcasing the first ever animated cartoon at a public exhibition. His animation, *Pauvre Pierrot*, or 'Poor Pete', contained 500 frames and ran for fifteen minutes. The audience gathered for the première at the Musée Grévin waxworks museum in Paris on 28 October 1892, but the world would have to wait for another thirty-six years before a certain little mouse sailed into the public imagination, and famously started it all.

7

HEALTH AND SAFETY, VICTORIAN STYLE

The Body Lightning Rod (1885)

As the nineteenth century drew to a close, an exciting new era began to dawn: one of fresh inventions and endless possibilities in all fields of science, medicine and technology. Edison had lit the blue touch paper with his electric wonders, but this new-fangled source of power held many Victorians in a grip of suspicion and trepidation, unsure of the potential jeopardies electricity could pose to their health. Though a general lack of awareness occasionally, and tragically, resulted in somebody being electrocuted in their own home, it was more usually the 'electric light men', or electrical engineers, who fell victim to shocking misfortunes. This was particularly true in America, where naked wires were often left exposed in the towns and cities, and there were innumerable reports of labourers being subjected to horrific electric burns.

In an effort to overcome this problem, an inventor named Mr Patrick Delane devised a 'body lightning rod' for wearing upon the person in order to prevent accidental electrocution. This personal safeguard consisted of copper cords that branched along the electrician's arms and legs, terminating at metal plates on the soles of the boots, thus reaching the earth. In the event of both feet being off the ground at once, a short length of metal chain or braid was allowed to drag on the floor. The idea was that if the engineer came into close contact with a surge of electricity, the bolt would strike the copper wire rather than the body, and be discharged to the ground in an instant without harming the engineer in any way. It was not recorded, however, whether or not the inventor had remembered to provide any sort of protective insulation around the copper.

Sunshades for Soldiers in the Desert (1885)

The following image illustrates a different sort of personal safety harness. In 1885 Britain found itself embroiled in the bitter Sudan

Campaign, part of a colonial war that was fought in the Middle East to put down the Mahdist Sudanese Army. While the British troops were abroad, the ladies of the National Aid Association were doing all they could to promote the comfort and welfare of the soldiers in the desert. Sunstroke was one of the most serious risks to which the men were exposed, and with a view to protect them from it, an effective new contrivance was designed. This was a kind of sunshade made from paper and bamboo, which was worn as a virtually weightless backpack. A piece of cane, bent into the shape of an arch, was fastened to each shoulder. To the centre of these arches were attached upright bamboo sticks, 18 inches high, which supported a canopy 24 inches long by 18 broad. This awning had a bamboo frame, which was covered with paper painted green on the inside, thereby shading the soldier from the blistering sun. Furthermore, this hands-free parasol conveniently permitted the serviceman to handle his rifle during battle.

The Greenhouse Heat Alarm (1884)

With the soldiers in Sudan suitably provided for, horticulturalists in Britain began demanding a solution to their own heat-related problem. Particularly during the hot summer months, people's greenhouses were prone to overheating. Once again it was the power of electricity that offered a solution, and a new electric apparatus was invented which automatically sounded an alarm if the temperature of a greenhouse rose to an undue level. As shown in the illustration, this new invention consisted of a voltaic cell, C, in circuit with an electric bell, B, decoratively shaped like an owl. Beside the owl was a similarly designed case, A, containing a mercurial thermometer, in which the rising of the mercury column to a given height completed the circuit between two wires that were fused into the mercury tube, thus causing the alarm bell to ring.

The success of this invention led to the development of similar heat alarms designed for use within homes, offices, and other buildings where an outbreak of fire would have had devastating consequences.

City Fire Alarms (1884)

In 1884, many decades before the majority of households could boast a telephone line, a system for quickly contacting the fire department was put into operation in America. Telegraphic fire alarms, which consisted of boxes attached to lamp posts or other convenient places, were installed around the towns and cities. When the boxes were opened, a mechanical device inside began tapping out a signal on the telegraph wire. This was sent directly to the local fire department, alerting them of the location of the box, and ensuring that the fire brigade left the station within a minute or so of the call-bell being sounded.

This process was by no means original, for the telegraph engineer Sir Charles Tilston Bright (1832–1888) had already introduced a similar system to the streets of Britain. In March 1883 his warning system was put into operation in the borough of Nottingham by the National Telephone Company, and saved many properties from the ravages of fire.

The Hand-Grenade Fire Extinguisher (1885)

A very quick and efficacious remedy for putting out smaller fires by oneself, such as a chimney that had caught fire, or a house fire that had not gained too general a hold, was reckoned to be the 'Harden Star Hand-Grenade'. This bizarre new invention consisted of a corrugated bottle of bluish glass containing a clear liquid, which the patentees claimed would not freeze and could be safely drunk with impunity or accidentally spilt on any kind of fabric without injuring it. The hermetically sealed bottle was only a few inches in diameter, and was easily taken in the hand by grasping it firmly by its neck. Thus held, it was simply thrown straight into the heart of the fire with a force sufficient to break the glass and liberate the contents. The moment the liquid came into contact with the fire it emitted large volumes of carbon dioxide, which suffocated the fire in a remarkably short space of time. In fact, it was claimed that this occurred almost instantly.

Glowing recommendations about the grenades stated that using them was as simple as throwing a stone, and advertisements asserted that they were perfectly safe, and in no way counter-beneficial to the health of the user whatsoever.

'I once saw one of these grenades extinguish a mass of flaming tar and petroleum in just a second or two,' enthused one testimony. 'A few hand-grenades would be useful articles to have about a house, provided the inhabitants, including the servants, were taught where to find them and how to use them.'

In the late 1800s the Harden Star Hand-Grenade was widely available for purchase from ironmongers, house decorators, and furnishers, and by 1885 they were used extensively across London, New York and Chicago. The London-based company that manufactured the grenades claimed that over 6,000 domestic

fires had already been extinguished by using the product, which cost just 45 shillings per dozen bottles.

By the turn of the century the hand grenades fell out of popularity, and it was later discovered that exposure to the 'risk-free' chemical – namely, carbon tetrachloride – that was contained within the glass bottles was seriously harmful to humans and animals. Furthermore, it was proved to be especially dangerous when exposed to high temperatures ... such as the naked flames of a house fire.

The Hand Fire Extinctor (1887)

In 1887 both the Marquis of Salisbury at Hatfield and the Birmingham Fire Brigade had reportedly adopted the use of a similar fire extinguisher termed the 'Lewis fire extinctor'. This was an invention about the size and shape of a policeman's truncheon. At the upper end was a wire loop by which it was hung on a wall from a nail or hook, and it contained a colourless, 'harmless' liquid comparable to the one found within the hand-grenade in that it was reputed to possess the same fire-stifling properties. In the case of an emergency, a sudden pull by the hand was all that was required to detach the extinctor from the cap, which opened the tube and drenched the blaze with the contents.

Experiments were carried out at the Crystal Palace in Hyde Park, London, with the aim of successfully extinguishing burning petroleum-soaked timbers. Ironically, despite the fact that the results were said to have been 'entirely successful', forty-nine years later this iconic glass building was completely destroyed by a terrible fire that ripped through the whole edifice in a matter of hours.

The Vulcan Water Sprinkler (1887)

By the mid-1880s automatic fire 'extinctors' and water sprinklers for mills and warehouses were gradually coming into general use, and one of these, termed the 'Vulcan', is illustrated in the engraving. In its ordinary form this apparatus consisted of a brass body fixed to the overhead water pipe of the room in which it was placed. The body had a tube projecting downwards at right angles, as shown, which communicated with the pipe. Over the orifice of this was fixed a cap, by means of sensitive solder, which melted at a comparatively low temperature such as would

be produced by an outbreak of a nearby fire. The melting of the solder released the cap and freed the water, which was projected by pressure into the neighbouring space.

Several varieties of the water sprinkler were made for different situations. The Vulcan opened at a temperature of 258° Fahrenheit, but there were other sprinklers of this order, notably the 'Grinnell', which opened at 289° Fahrenheit, and the 'Simplex', which opened at 360° Fahrenheit. In general a sensitive sprinkler – that is to say, one opening at a comparatively low temperature – was preferable, since it did not allow the fire to gather strength before it operated.

The Grinnell Fire Sprinkler (1885)

The following diagram explains the action of the Grinnell fire sprinkler, which, by 1885, was being largely used across America. It was the eponymous invention of a Mr Grinnell of Rhode Island, and the remarkable success of the sprinkler led to its gradual introduction to buildings across Great Britain. This fire-extinguishing system consisted of lines of small pipes carried throughout a property, which were connected to the public water

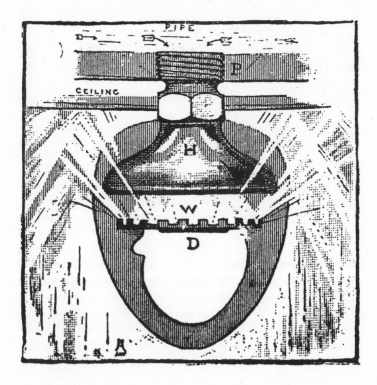

main or water tank at the top of the building. The pipes ran along the ceiling, and every 8 or 10 feet sprinklers were attached to them. In the diagram, P represents the screw that attached the device to the pipe. The extinguisher, as can be seen, consisted of a heart-shaped envelope enclosing a kind of bell-mouth, which opened by a hole into the pipe. This hole was ordinarily closed by a diaphragm, D, held in place by a trigger, or lock, fixed by fusible metal. When 289° Fahrenheit was reached, owing to the heated air of the fire rising to the ceiling, the fusible trigger melted, the diaphragm fell a small distance into a niche provided for it, and the water, under pressure, escaped as shown at W. On striking the diaphragm, which was toothed at the edges in order to distribute the water in jets, spray was deflected back up to the ceiling, from which it fell in a regular shower.

The advantage of such an appliance was that it operated automatically, and, fundamentally, before the fire had attained serious proportions. Even the electric fire alarms hitherto discussed were open to the objection that they did not bring immediate help, but merely summoned attention to the fact that a building was potentially at risk. In the case of the Vulcan and Grinnell extinguishers, a means of extinguishing the flames was already at hand, and for public theatres, cotton factories and other such places, the invention was certainly commended.

Fire Exits (1887)

It was 25 May 1887 when a devastating fire ripped through the Opéra Comique theatre in Paris, resulting in the death of eighty-four people, including firemen and police officers who had tried to assist. Hundreds more were injured and trapped in the debris, which took days to clear, and on 4 June the *Manchester Courier and Lancashire General Advertiser* ran a very sad story documenting the aftermath.

'Since the morning of the fire,' the paper reported, 'an old gentlemen, looking very pale and sad, has been sitting on a fragment of beam opposite the entrance through which the dead bodies are brought to be put in coffins and carried away. He went with his wife and daughter to see *Mignon* acted. He was separated from them on leaving the hall, but supposed they had gone out before him. They have not been seen since, and he has taken up his post on the log, where he sits with his eyes fixed on the spot to which the dead bodies are brought to be borne out. Every time a new one comes he rises, looks at it, shakes his head, and returns to his seat sobbing.'

It was said that this terrible tragedy, which was widely publicised throughout Europe, could have been avoided if special 'exceptional exits' had been installed in this old-fashioned and dangerous building, which was completely destroyed by the blaze. It was common practice for nineteenth-century theatres to be fitted with reservoirs of water for flooding the stage in case of a fire, but unfortunately not at the Opéra Comique.

The severity of the 1887 disaster gave rise to many proposals aimed at preventing further loss of life during times of great panic in public buildings. A gentleman named Mr B. M. Norwood offered one such suggestion, recommending that, firstly, all available exits should never be kept locked, as was the case in Paris. Secondly, he proposed that 'fire exits', for use purely during times of emergency, be installed in every public building. Partitions of wooden plank, thick enough to withstand moderate pressure on ordinary occasions, yet thin enough to break down easily under force, could have been fitted over these specially prepared emergency exits. Mr Norwood envisaged that these planks might be inscribed with the notice, 'In case of panic, break through.'

Another submission advocated the employment of fire attendants permanently stationed at the exit doors ready to unlock them at a moment's notice thus allowing the assembled public to spill outside, should any panic ensue. A third suggestion, which came from Germany, had it that that luminous paint should be used to mark the exits in case of a blackout.

Over in Brussels, Monsieur Leon Lenaerts devised an electronic system whereby upon the outbreak of fire in a theatre or concert hall, any member of the audience could turn off the gas at the meter by means of an accessible switch. This simultaneously lit a series of electric lamps, which illuminated the exits. It was expected that the assurance of this arrangement would give confidence to an audience, thus preventing a mass panic.

The Portable Saddle Fire Escape (1884)

In the days before modern health-and-safety legislation was even a twinkle in Parliament's eye, an emergency exit was a relatively novel concept for the Victorians. However, with domestic gas and electrical installations on the increase, a nationwide demand for the introduction of emergency escape plans in public buildings was high, prompting inventors to start imagining new safety contraptions to assuage the populace.

The figures illustrate the use of a new portable fire escape which could be wound up and carried around upon one's person

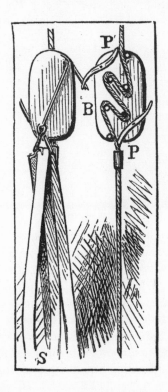

in case they found themselves in a life-or-death situation. This 'saddle fire escape' was invented by two American gentlemen named Messrs D. Ware and C. W. Richman, of Philadelphia. Their device consisted of a long rope with a leather loop, or saddle, S, at one end. The saddle was situated below a wooden board, B, which had six pegs, P P, on it, around which the rope was twisted in the manner shown on the previous page. Two friction brakes were also brought to bear on the rope at either end of the board. The object of this board was to keep the person from plunging too quickly towards the ground, the friction pins and brakes rubbing on the surface of the rope and slowing the speed of descent.

To use the escape, a hook was caught in the windowsill to secure the rope, and one end was thrown to the ground. The escapee was then required to climb into the saddle and step off the window ledge, into the air. By controlling the brakes on the board with the hands, the person could safely make their descent down the side of the building, thus escaping danger, in the manner illustrated opposite. The friction brakes were so powerful that one could not only moderate the speed, but also stop on the way down, if required.

On reaching the ground, the rope, with its saddle and board, could be drawn back up again ready to be used by the next person requiring to be saved.

The Ladder Fire Escape (1887)

In 1887 another kind of small and handy fire escape was introduced, which was described as being particularly useful in

countryside dwellings where public services for rescue were not close at hand. The apparatus could be packed away so compactly in its case that its permanent presence in a room was said to cause no inconvenience to the inhabitants whatsoever. It was to be kept fixed below the bottom sash of the window by means of the two metallic cords as shown on page 183, and when required for use it was only necessary to lift the ladder and drop it from the open window, as shown opposite. The stays shown in the engravings served to prop the ladder from the outer wall of the building.

The ladder was advertised as being not only rustproof, but also – as one might hope – fireproof as well, and was strong enough to bear the weight of several people at any given time.

The Bucket Fire Escape (1884)

Many further proposals for household emergency escapes were conceived in the final two decades of the nineteenth century, but for a variety of reasons, including cost, danger, and cumbersomeness, a great proportion of them did not long survive their inauguration. However, in 1884 a London merchant named Mr Thomas Hale invented a new type of window fire escape that attracted some notice. The escape itself consisted of a large bucket, made of a combination of wood and strong sailcloth, which was spacious enough to hold two adults or three to four children. The bucket was light enough to be easily carried and placed in position by just one person, and it could be fixed to the window ready for use in less than a minute.

The supporting apparatus comprised two upright iron rods, which rested upon the outside of the windowsill, and which were kept in position by a parallel bar placed against the inside of the window frame. The rope to which the bucket was fastened ran through a pulley attached to the rods. Once the occupants of the building had clambered inside the bucket, the apparatus could be let down from either inside or outside the window, and one person could lower the bucket down to the ground unaided and without danger.

The Pocket Ambulance (1885)

This general increase in awareness of public safety led to the development of a 'pocket ambulance'. A London-based firm prepared a little book intended for the use of people who had learned to 'render early aid to the wounded' through the teaching of the St John's Ambulance Association. The book was to be carried around on one's person at all times, as it contained a number of practical articles on how to deal with an emergency – such as how to stop a person's wound from bleeding – with helpful step-by-step instructions.

The St John's Ambulance charity still produces its pocket ambulance in the form of a handy first-aid guide that can be ordered, free of charge, through their website.

The Accident Ambulance (1884)

Although the Victorians were aware of the importance of being able to deliver basic first aid to the needy, it was often necessary to convey the more seriously injured casualties to hospital, where professional medical attention could be received. Emergency vehicles were first employed as early as 1487, when the Spanish developed a mode of transportation to move injured soldiers across the battlefield. However, civilian alternatives were not

brought out until the 1830s, and these generally involved the use of a horse and some kind of cart. By 1884 Messrs Atkinson and Philipson, who ran a coach-making business in Newcastle-upon-Tyne, had designed a new type of ambulance 'for the cure of persons suffering from accident'.

The top image on the previous page represents their emergency waggon. The axles were fitted with India-rubber collars to prevent noise and jarring. The springs were strong but elastic; the step at the back was long and broad; and the doors opened outwards. In the floor of the car was a grooved track to hold the wheels of the stretcher shown in the bottom figure. This was a convenience in itself, and consisted of an ash frame holding a spring mattress and air pillow. A pair of light wheels with rubber tyres were attached, to enable the stretcher to be run along the floor of a hospital with speed. The wheels were, however, detachable at will.

For the provision of more than one patient, it was also possible to suspend a second stretcher from the roof of the ambulance. This was made of canvas on parallel poles, which could be slipped out when the invalid was placed upon a bed without disturbing them – a great advantage when a patient could not bear movement. The stretcher was hung by rubber hooks from the roof, and was kept in place by crosspieces.

As well as room for the stretchers, there were also seats provided inside the ambulance for three attendants, which, with two invalids, the driver and his assistant, made seven passengers in total. The vehicle was so designed that one horse could easily draw the load, and the car was well lit and ventilated.

The Bicycle Ambulance (1887)

Around three years later it was proposed to phase out such waggon-based ambulances and replace them with pushbikes instead. These new bicycle ambulances were built by joining two cycles together with a pole, thus keeping the wheels apart, and uniting both bicycles into one vehicle. A hammock was then slung from the seats of the two bicycles by means of crosspieces and hooks, on to which the patient could be 'securely' laid.

Despite an initial interest in the vehicle, the bicycle ambulance was scrapped just a few years after the idea was conceived owing to the invention of the automobile, which inevitably took precedence in the development of emergency transport. Nevertheless, since the year 2000 bicycle ambulances have once again been spotted darting up and down certain streets of Britain. This 'Cycle Response Unit' was formed by the NHS in an effort to swiftly transport paramedics through the most congested thoroughfares.

The Ambulance Steamer (1884)

In Victorian times, poverty-stricken inhabitants of the most deprived parts of the country were particularly vulnerable to any number of horrific diseases, yet the majority could ill-afford the expense of a doctor. With the welfare of these desperate paupers in mind, a London organisation known as the Metropolitan Asylums Board was set up in 1867 to care for the capital's increasing number of ailing indigents. The board established a number of purpose-built state institutions where the sick, who would otherwise have been left to linger in the squalid conditions of the dreaded workhouses, could be treated.

By the 1880s the MAB had also begun to reform London's ambulance service, which, the newspapers reported, was in a most shocking state indeed. When the board was first set up, emergency vehicles for transporting the poor were arranged by the Poor Law Guardians, though very few provisions for the care of the patients were made. It was reported on 30 September 1887 in the *London Daily News* that in November 1881 one such horse-drawn ambulance arrived at Deptford Hospital carrying two patients suffering from severe cases of smallpox. One was a woman, and the other her four-month-old baby.

'The driver was drunk,' the newspaper criticised, 'and had galloped nearly all the way, the door of the vehicle being at one time jolted open and the smallpox patients nearly thrown out. When the ambulance reached the hospital, the driver, in attempting to get down, stumbled and broke his leg, and was driven off to the workhouse in the vehicle from which the smallpox patients had just been discharged.'

The article stated that when no ambulances were available in cases of emergency, 'public vehicles were frequently employed in convening smallpox and fever patients [to hospital], and drivers in charge of such "fares" have been often known to leave them standing outside a public-house while they, the drivers, have gone in to get drunk.'

The MAB was determined to put an end to this kind of misman-agement, and in 1884 they set up their own ambulance service in the form of a special paddle steamer on the River Thames. Not only did this make it rather inconvenient for the driver of such a vehicle to pull up outside the city's public houses, the vessel was also fitted with all the modern amenities for the treatment and transportation of invalids. The steamer was 105 feet long, 16 feet 6 inches in beam and 6 feet 6 inches deep. The hull was made from iron, the deck of yellow pine, and the fittings of teak. She was divided into six watertight compartments; forward of the machinery was the hospital, which was partitioned by a bulkhead to separate the male and female inmates. The hospitals were fitted with iron beds, heated by hot-water pipes, and provided with medicine closets, doctors' and nurses' rooms, and a whole host of essential commodities, while aft of the machinery was the convalescent hospital.

The Floating Hospital (1887)

In 1881 Britain was gripped by fear, for the fifth cholera pandemic of the century was reckoned to be imminent. Eager to assuage the growing panic, the River Tyne Port Sanitary Authority in the north-east of England began work on a floating hospital on the Jarrow Slake mudflats in the River Don, south of the Tyne. The idea was not only to treat those affected, but also to try to lessen the threat of the disease spreading to the rest of the population by separating the hospital from the mainland by way of a significant stretch of water.

The accompanying figure illustrates the hospital. It was floated

on ten cylindrical iron pontoons, each 70 feet long and 6 feet in diameter. These were detachable for inspection and cleaning purposes, and were supported by a foundation of iron girders on which the construction was raised. The hospital consisted of three main wooden buildings and several additional smaller ones, staffed by all the necessary auxiliaries.

It transpired that the pandemic did not spread as far as Britain, but the floating hospital was opened anyway in 1886, and remained in use until 1930, when it was dismantled and towed away for scrap metal.

The Microphone in Medicine (1893)

Though advancements in science were forcing a gradual progression in the field of medicine, the Victorians lived in a world where many treatments were still based on outdated botanical recipes, and the ancient practice of bloodletting, for instance, remained a common 'cure' for many ills. As the century neared its end, however, new forms of medical treatments and scientific equipment were developing apace, and one noteworthy breakthrough was the adoption of the microphone to detect sounds too feeble for the unaided ear. The eminent scientist Professor David Edward Hughes (1831–1900) had invented this auditory device in the 1870s, which some years later was fitted inside the ordinary stethoscope. However, owing to its delicacy and want of electrical knowledge on the part of the doctors, its use in auscultation did not become general until around 1893. This was partly in consequence of the shocking case of a Russian lady who was saved from premature burial by means of a microphone placed over the region of her heart, which could clearly be heard beating, even though she had been considered quite dead. A New York physician named Doctor Pleydell was thus inspired to invent the 'micro-stethoscope', by which he could distinguish sounds of the heart, lungs, blood vessels, and other parts of the body, which were hitherto wholly inaudible to the ear alone.

The Hearing Test (1885)

Scientists and inventors of the nineteenth century soon began offering medical professionals across the western world a range of general health-testing instruments, but one for accurately testing a person's hearing was not introduced until Professor Alexander Graham Bell (1847–1922), the patentee of the telephone, devised an instrument specially adapted for this purpose. It consisted of a telephone and an arrangement of two coils with a battery and a rotating current interrupter, which caused the telephone to emit a single musical note. The test subject was required to hold the telephone to their ear, and the volume of the musical note was gradually diminished. This diminution was achieved by slowly drawing the coils apart, and the distance between the two, as indicated on a scale, was a measure of the intensity of the sound. When the person could no longer hear the note, the number on the scale gave the acuteness of their hearing.

Professor Bell obtained some useful results with the apparatus. He discovered that around 10 per cent of children from New York had slight defects of hearing, and about 1 per cent of those tested were found to be completely deaf.

The Koniscope (1893)

Meteorologist and physicist Dr John Aitken F.R.S. (1839–1919) spent his working days meticulously studying atmospheric dust. In 1893 he brought out a little apparatus known as the 'koniscope' for determining the sanitary condition of the air in a given room. A small air pump, shown at P in the figure, sucked a sample of the air through the stopcock, S, into a test tube, T. This had glass ends, E E, and was lined with vitreous paper. The tube was then turned to the light and peered through like a spyglass. The colour of the enclosed air, when compared to a scientific scale, was an indication of its purity.

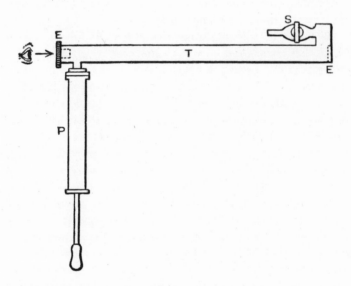

The Menthol Inhaler (1887)

Airborne diseases were commonplace during Victorian times, when good sanitation was often lacking. Those who lived in the capital city and other large metropolises often found that fresh air was a luxury that could only be found in the countryside, far from the smoking chimneys and smoggy streets. This urban

uncleanliness invariably led to a variety of respiratory problems, including diphtheria, TB, and bronchitis.

Owing to its soothing properties, the Victorians were well aware that menthol was a wholesome tonic which helped ease ailments of the throat and lungs. It was also believed to reduce the severity of headaches, and lessen the effects of flu, hay fever, and many other common complaints. Consequently a new type of inhaler, as illustrated in the engraving, was brought out that utilised this marvellous herbal remedy. The device consisted of a small glass tube, narrowed at one end, and provided at each extremity with a closely fitting stopper. The centre of the tube was filled with loose crystals of menthol, secured by a perforated cork at each end. The great convenience of such a compact device was that sufferers could carry it around during the day.

In cases of a head cold, or other head troubles, both stoppers were removed from the inhaler, and the narrow end was inserted into one nostril. The person's breathing was then confined through that nostril only. The mentholised air thus inhaled was quickly carried to every part of the head, and instant relief was afforded. When the throat or lungs were affected, placing the broad end of the apparatus between the lips and drawing long, full breaths through the inhaler was said to speedily afford relief to the troubled areas.

The Bronchitis Tube and Kettle (1892)

Many Victorian homes possessed a 'sick room' where ailing members of the family could be confined for treatment and recuperation. A new utensil, introduced for use in these rooms, was the bronchitis tube, which could be readily attached to the top of any ordinary kettle and so convert it into a 'bronchitis kettle'. It was believed that a sufferer of this disease, or any other chest infection, would find it easier to breathe if the air in the room was moist, and a bronchitis kettle was used to achieve this. It was filled with water, with a drop of menthol added to ease congestion, and left to simmer over a fireplace. Steam began flowing out of the long bronchitis tube and was dispersed through the air.

The method of attachment was very simple. A number of discs of different sizes were supplied with the tube, and the one which better corresponded with the aperture of the kettle lid was selected. This was held in place by a sliding loop of wire, on which was fitted the long tube.

The Domestic Ozoniser (1893)

Thanks in part to the studies of the notable microbiologist Monsieur Louis Pasteur (1822–1895), the Victorian population gradually became conscious of the existence of germs, and specifically the fact that via these microorganisms diseases could be passed from one person to another. Scientific inventors subsequently became engaged in a relentless search for new and improved ways to understand and eliminate these dangers from the world, and it was widely agreed that one of the main ways of preventing infection was to keep one's environment clean.

Since the start of the Industrial Revolution, when vast amounts of fossil fuels were burnt in order to power mankind's new machines, levels of so-called 'greenhouse gases' in the earth's atmosphere have risen by up to 40 per cent, according to experts, and they continue to increase. Ozone is one such gas, produced by the toxic emissions that are pumped out of vehicle exhausts

A·A

every single day, creating a dangerous ground-level ozone layer across twenty-first century Britain. Studies have shown this can seriously affect the respiratory system of anyone exposed to high levels of this pollutant, triggering conditions such as asthma and bronchitis.

Ozone is also naturally produced in the earth's atmosphere by the electric discharges generated during a thunderstorm. Victorian researchers of this phenomenon could not fail to notice that once a storm had passed, a fresh fragrance was left hanging in the ether, leading them to conclude that ozone must be a natural purifier of the air. After much examination it was discovered that, under laboratory conditions, the electrical discharges from an alternating current could be employed for manufacturing this natural sanitising substance on a commercial scale.

It was first necessary to have sparking points a little apart from each other, and to connect these in circuit with the current. The sparks passing from one set of points to the other through the gap of air between them mimicked a thunderstorm and produced ozone, which, when circulated around a room, was found to be a healthy disinfectant for one's home.

The engraving shows an 'ozoniser', or 'ozone maker', for household use. It was devised by an inventor named Sig. Andreoli, and was shaped like a music stand. The quadrangular apparatus consisted of a sheet of glass covered with tinfoil, having a series of toothed iron strips, S, above it. The strips and tinfoil were connected to the electric circuit by the wires, W W, and the discharge passed between them, thus producing the ozone, which was then free to flow about the residence.

The Disinfecting Lamp (1885)

In France, a committee of experts, including Monsieur Pasteur, conducted a series of experiments in order to ascertain the best means of disinfecting chambers in which cases of contagious infections had been lodged. The committee reported that sulphurous acid gas was the best disinfectant, but recommended that instead of simply burning sulphur, as was done in Victorian Army barracks and other such places, bisulphide of carbon, also known as carbon disulphide, should be burned in the rooms of homes, as it was deemed less injurious to the furniture. A lamp, as shown in the illustration, was therefore devised for enabling this volatile chemical to be burned with safety, in the hope of decontaminating a sick room.

The disinfecting lamp was made of copper, and consisted of an outside vessel, A B C D, containing the lamp, I H E F. Three bent copper tubes, R S, passed through the sides of the lamp, and

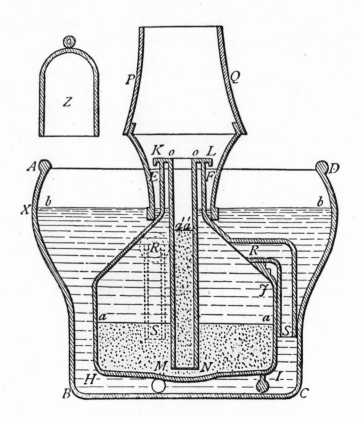

the cylindrical tube, K L M N, containing a cotton wick, reached from the top to the bottom. A copper chimney, P Q, surrounded the flame.

The lamp was filled with bisulphide of carbon to about the level *a a*, and the outer vessel was filled with water to about the level *b b*. By means of the bent tubes, the water passed into the interior of the lamp and forced the bisulphide to enter the copper tube, up to the level *á á*, where it was absorbed by the wick and ignited at the top of the tube, *o o*.

As the chemical burned away and was released into the atmosphere, the water took the place of the bisulphide, thus allowing the lamp to extinguish itself. The combustion could be regulated by adding more or less water, so as to raise or lower the level *b b*. After several hours of burning, the room was considered sterile.

What this French committee of experts failed to recognise was the fact that inhalation of carbon disulphide could trigger any number of adverse health effects, including chest and muscular pains, loss of memory, numbness to the hands and feet, impairment of the nervous system, infertility, damage to the heart and internal organs, and even death.

The Disinfecting Oven (1885)

In addition to the lamp, a disinfecting oven was also invented for domestic use. The accompanying woodcut offers a cross-section of this new device, which was designed for sterilising contaminated articles of apparel and bedding. The oven was fitted

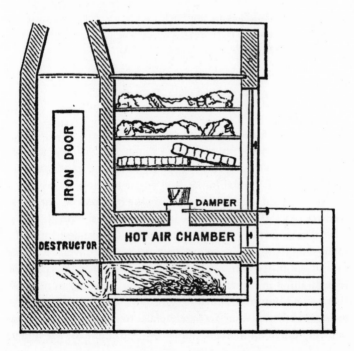

with airtight iron doors, with shelves inside to hold the goods to be fumigated. Inside the oven was a furnace with a hot air chamber above. The top of this chamber was made of iron, and an aperture in the middle was fitted with a damper, which was drawn from the front to regulate the heat. A little window was placed in the wall of the front to show a thermometer within. When the temperature was too high, the damper was pushed in, and vice versa.

The disinfectant was prepared by placing a small bucket containing a pint of hydrochloric acid above the damper, and throwing a handful of highly explosive potassium chlorate crystals into the acid. The door was then firmly shut, and the temperature raised to the required degree. A chemical reaction then took place inside the oven, whereby chlorine, a useful disinfectant, was released into the air. This circulated around the apparels, and after three hours the goods were considered wholesome.

For contaminated articles that were beyond salvation, a 'destructor' was added to the back of the oven for burning up items that were to be entirely destroyed, such as the wallpaper of an infected chamber.

Swan's Electric Safety Lamp (1887)

Explosive substances in confined spaces posed serious risks, and an encounter with a cloud of flammable gas was the last thing Victorian coal miners wished for. Firedamp was a terrifying danger down the mines, and engineers were under much pressure to find an effective way in which to swiftly and effectively detect this highly volatile gas before it was too late.

In 1886, at the 56th British Science Association meeting at Birmingham, Sir Joseph Wilson Swan, the inventor of the incandescent light bulb (whose house in Gateshead became the first dwelling in the world to be lit by such means), announced that he had improved his design for the miner's safety lamp. In essence, he had invented a hand-held battery-powered lamp that not only provided a fundamental source of light for the miners

while they were underground, but was also capable of indicating the presence of firedamp.

Swan's new electric battery consisted of four cells, and emitted a light equal to one and a quarter candles, which lasted ten hours before going out. The teak box was liquid-tight so that the lamp could be placed in any position underground. By turning a switch, the current from the battery was sent through a fine platinum wire contained inside a glass tube underneath the lamp, as shown. If firedamp was present in the atmosphere, the wire began to glow abnormally bright, thereby providing fair warning to all the colliers in that part of the mine.

The Firedamp Detector (1885)

The mining engineer and Yorkshire colliery proprietor Sir William Edward Garforth (c. 1846–1921) introduced his own simple test for identifying the presence of deadly firedamp. A small, hollow rubber ball, such as those used by Victorian children, had a hole bored into it, and was taken deep into the mine by the collier. When he wished to test the ether, he simply squeezed the ball and released it again, thus filling the ball with air. He then repeated this process, allowing the jet of escaping air to play upon the flame of his lamp. If the dreaded 'blue cap' appeared on the flame, and the flame lengthened out, this was a clear indication that the atmosphere contained firedamp, and it was time to get out.

The Life Preserver at Sea (1884)

Despite the many subterranean dangers, it was argued that there was no place on earth more perilous than out on the open waves. Goods and merchandise were regularly transported overseas, and many a freak accident occurred when a vessel was broken up and consumed by the clashing tides of a storm.

One of the chief elements of danger to a good swimmer cast adrift in a rough sea lay in the suffocating and stupefying effects of the waves that broke over their head. Even when the body was suspended by a lifebelt, drowning frequently resulted from the sheer force of the ocean. To provide protection against this, a gentleman named Mr William Wilkins invented a new life

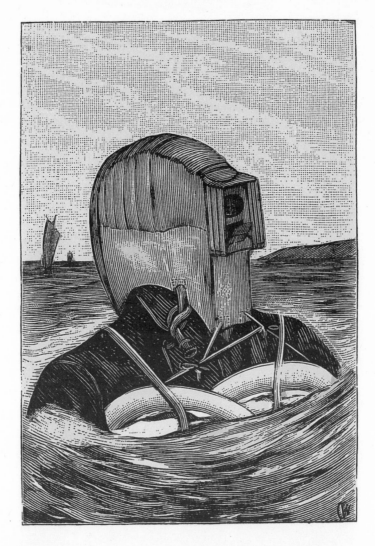

preserver consisting of an atmospheric helmet and belt. The former was provided with ample arrangements for sight, speech, and, quite crucially, respiration. Both the helmet and belt were divided into two compartments, offering buoyancy to prevent the wearer from sinking. Indeed, it was calculated that the weight of a fully dressed man in the water amounted to only 6 lb, but the belt could sustain up to 40 lb, and the helmet 20 lb. In case of an emergency at sea, both helmet and belt could be inflated and put on in about three minutes.

This new life preserver was exhibited in London at the Great International Fisheries Exhibition of 1883. In a trial in a rough sea off Brighton its efficiency was fully demonstrated, and it seemed likely that the invention would be found particularly useful to members of lifeboat crews, who faced the danger of being washed out of their boats and drowned before completing their rescue mission.

The Diver's Helmet (1887)

This image illustrates a similar state-of-the-art deep-sea diver's helmet, which was fitted from above with an electric incandescent lamp, capable of directing a 50-candlepower beam under the water. As well as the obligatory breathing tubes, a telephone

was also fixed to the inside of the helmet, which enabled the diver to speak with his assistant on board the boat.

The Life-Saving Pillow (1893)

For the safety and convenience of passengers aboard luxury Victorian liners, a practical lifebelt was developed, which was always at hand should the dreaded call be made to abandon ship. The berth pillows provided in each cabin were not just ordinary berth pillows: they were live-saving pillows. By 1893 these novel lifebelts had been adopted on board various vessels, including the Cunard steamers *Campania* and *Lucania*.

The following two images show the double-pillow design and the mode of attaching it to the person. Each pillow contained

a series of air chambers covered with a stuffing of hair and feathers to make a soft headrest, and the buoyancy of the belt was three times the minimum required by the Board of Trade. As the belt was easy to put on and even easier to locate, it was hoped its implementation would inspire confidence in even the most nervous passenger.

The Whitby Lifebuoy (1885)

The 'Whitby' lifebuoy was another safety device that could be found on board Victorian ships. It was well equipped for life saving, and had been supplied to the Navy by 1885. As illustrated, it consisted of a cooper buoy divided into eight watertight compartments, and fitted with a chain on which to rest the feet in the manner shown. Two flags, a whistle, and two hand-lights for use in the dark were also provided, as well as a handy flask for spirits, should the individual in peril require a warming, alcoholic infusion while they awaited rescue. Moreover, two calcium lights were also attached to this luxury lifebuoy, which burned, on contact with the water, for over an hour.

The Trunk Lifebuoy (1884)

The main objection to existing life-saving devices was the fact that no provisions were made for the salvage of one's belongings. With this in mind, the trunk shown in the following

engraving was designed to not only save lives at sea, but also as many treasured possessions as possible. The crucial characteristic of the trunk was that its lid could be opened to reveal the hollow interior, which was divided into useful compartments for storing personal valuables while the unfortunate castaway awaited rescue.

The trunk was made of light wood or cork, covered with waterproofing, and a wing of cork was fitted along the bottom edges to further increase the buoyancy of the device. An attached flagpole accommodated a flag that could be hoisted to signal distress and attract the attention of any passing ships. On the top of the buoy were rows of eyes into which a rope could be slung for holding on to, or for coupling two or more trunks together to form a raft.

The Treasure Safe (1884)

A secure repository to keep one's treasures in was a communal desire among all the social classes, but particularly amid those who had amassed a sizeable collection of valuables. In the late nineteenth century the Vanderbilts were one of America's wealthiest families. Originally hailing from the Netherlands, the dynasty chose to lay down roots in New York in the 1650s. Over the centuries the family's riches grew, and by 1877 the railroad magnate Mr William Henry Vanderbilt (1821–1885) had inherited a fortune of $100 million from his late father's estate.

Mr Vanderbilt's custom-built money-coffer was a strong box indeed. It was designed to be fire-, water-, and burglar-proof, and it was certainly a place where vast sums could be safely stowed away without worry. In fact, Mr Vanderbilt's magnificent treasure safe, which measured 36 feet by 42 feet, was said to be the most impregnable fortification on the American continent. Its foundations were blasted out of rock, the front wall was 5 feet thick, and the side and rear walls, made from pressed brick and brown stone, were 3 feet thick. The beams, girders, and main pillars were made of iron, encased in fireproof material, and the doors, window frames, and partitions comprised iron, marble and glass. No wood entered the structure, for this was easily burned. Each of its outer doors weighed 8,200 lb, and were fitted with the most approved modern locks and alarms available to man.

William Vanderbilt's luxurious estate, situated near Firth Avenue, New York City, no longer exists, but his son, Mr Cornelius Vanderbilt (1843–1899), constructed a lavish summer home on Rhode Island called The Breakers, which still stands. Like his father's notable treasure safe, The Breakers was designed to be completely secure from intruders and natural disasters. Today this opulent mansion receives 300,000 visitors every year and remains one of America's most famous National Historic Landmarks: a lasting legacy to the fortitude of one safety-conscious family who lived and thrived in the bygone Gilded Age.

8

SCIENCE AND NATURE

Projecting Microscopic Views (1887)

Science was burgeoning towards the end of the Victorian era. Many local authorities had invested considerable sums in new museums, libraries, lecture theatres, and other educational establishments in an attempt to better the national reputation of the great industrial cities, which had seen a drastic increase of impoverished slum areas. Fashionable exhibitions and public displays were regularly organised with a view to attracting crowds and proving the sceptics wrong about these large urban settlements. As a result, learning became accessible to a wider audience, and improvements in modern technology meant that knowledge could be shared with a far greater ease than in previous centuries.

Photography, for example, had opened a hitherto unknown door into the world of science, enabling researchers to examine their studies more closely. In 1887 Professor Salomon Stricker (1834–1898), an experimental pathologist from Vienna, succeeded in projecting, on to a white screen, images of objects as seen through the microscope, magnified from 6 to 8,000 times. This was achieved by means of an arc lamp of 4,000-candlepower, enclosed in a camera that had powerful magnifying lenses. The light was directed through the microscopic bodies, which were placed on glass slides, and by means of the lenses, their images were projected on to the white screen. In this way it was possible to teach a whole class the secrets of anatomical structure, as disclosed by the microscope.

The Great Equatorial Parisian Telescope (1884)

Over in the fashionable city of Paris, a great seventeenth-century observatory was fitted with a very fine equatorial telescope, as

shown in the illustrations taken from the *Scientific American* journal, published on 23 May 1891. The telescope had been designed by the astronomer Monsieur Maurice Loewy (1833–1907). Its chief peculiarity was that the telescope was bent in the middle to an angle of 90°, the whole thus forming two sides of a square. The light from the night sky was reflected round the corner and up to the eyepiece by a mirror at the bend. The viewing end of the tube rose upward, and the observer was able to sit comfortably in a chair above the telescope, looking down into the reflected sky as if examining a microscopic specimen. This arrangement permitted a greater ease of working, and the usual cumbrous dome over the instrument could be dispensed with.

The Pocket Anemometer (1885)

While nineteenth-century astronomers were gazing at the stars, new technology was being developed for those who studied earth-based physics. By 1885 a handheld anemometer for measuring the speed of the wind was invented by Sir Francis Galton F.R.S. (1822–1911), the cousin of Mr Charles Darwin (1809–1882). The device consisted of Robinson cups – commonly found on old-fashioned anemometers – and a dial indicator, which gave the velocity of the wind in miles per hour, the time being taken by a sandglass. To take a measurement, the dial was allowed to record until the sand ran out. The very instant this occurred, the dial was thrown out of gear by the shaft of the cups, and the number of miles read off.

The Tintometer (1887)

The 'tintometer' was the name of another new scientific machine, which was, as the name suggests, designed to measure the depth of colour in liquids and solids. In particular it was used by brewers to ensure the highest quality and colouration of beer. The contraption consisted of two tubes 18 inches long, open at their ends, and placed side by side. In one tube was placed the liquid to be examined, and in the other a series of coloured glass, arranged in a scale of colour, with which the comparison of colours was to be made. Both of the tubes were viewed through a single eyepiece so that the precise tint could be determined.

The Electrical Ore Finder (1893)

By 1893 the latest application of electricity had been made by an American inventor of some distinction, Mr George

Milton Hopkins (1842–1902), for the detection of underground metallic ores. His 'electrical ore finder' consisted essentially of a coil of insulated wire in circuit with a battery and an interrupter, which were contained in a box that was carried over the prospector's shoulder. At an angle to this coil was another wire, in circuit with a telephone. The prospector was required to hold the telephone to his ear while suspending the coils over the ground. If the ground was free of any magnetic ores, the telephone remained silent, but if metal was present in the soil, it would be detected by the electrical ore finder due to a magnetic interruption in the electrical current, and would cause the telephone to ring.

The apparatus was especially useful when the nuggets or minerals were near the surface of the ground. When searching for metallic ores underwater, the coils could be enclosed in a watertight case and used in much the same way.

The Gold Finder (1892)

Alluvial gold diggings, that is to say, those in which the precious metal was discovered in nuggets under the soil, were most remunerative to the prospector who had not the time nor the wherewithal to embark on the laborious task of quartz-crushing in his hunt for riches. In order to aid the prospector in the discovery of gold at a certain depth below the surface, the electric probe, or 'electric gold finder', here illustrated, was invented. It consisted of a steel boring tube that contained a metal rod connected to a battery and an electric bell, which were carried separately by the prospector. If the point of the probe struck against a nugget of gold underneath the soil, the electric circuit was established and the bell rang. A sample of the soil thus found to be auriferous could be obtained by advancing the outer tube into the ground beyond the inner core, and scooping up an amount for analysis.

The Earthquake Recorder (1884)

Like many fields of study, earth science was an emergent subject for the Victorians. Geophysics was one such area of interest, and in 1884 a Scottish inventor named Professor J. A. Ewing, of Dundee, devised the apparatus illustrated in the following image for measuring earthquake shocks. His invention was adopted for this purpose in Japan, where significant seismic activity was commonplace.

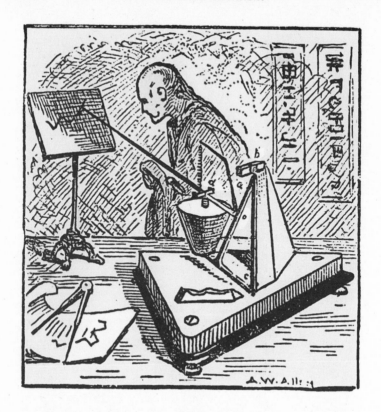

The professor's device consisted of a light steel triangle, *a*, which carried a metal rod, pivoted at *d* on a vertical axis which passed through the centre of percussion of the triangle. A tracer – which served to magnify as well as to record the movement of the earth – was made from a long straw tipped with steel. This was attached to the rod by a horizontal joint at *d*, allowing the tracer to accommodate itself to any slight inequalities in the surface of the glass plate on which its distant end rested, with a weight sufficient enough to create a measurable line on the surface. Any tremors from the earth were detectable by the earthquake recorder, causing it to vibrate. In turn, the vibrations would travel up the rod and across the tracer, thereby agitating the steel tip. As the glass plate was coated in a sooty substance, a clean line was drawn upon the glass by the steel end of the tracer as it scratched through the soot. The line thus drawn by the point on the sooty surface was a true record of the earth tremor.

The Brachionigraph (1887)

Victorian physicians found that modern technology was offering them novel solutions to medical quandaries that centuries or even decades earlier would have been considered untreatable. The brachionigraph was one such invention, developed for use around 1887. This ingenious instrument was designed to enable people who had lost the use of their hand, or indeed the entire hand itself, to write again.

Simple in construction, it consisted of a light metal strip, shaped to fit the side of the forearm, and supported by a kind of gauntlet that was attached around the arm. The brachionigraph carried a pen at its extremity, and the person could move the pen using the muscles of their arm and shoulder, as if writing normally.

The Hypnoscope (1885)

Though it was not an exact science, another device was constructed to fit on to a person's hand which could allegedly test whether or not that person was susceptible to mesmeric influence, or hypnotism. The 'hypnoscope', still used by some hypnotists today, was a little gadget that was created by Dr Julian Ochorowicz (1850–1917), a Polish inventor and psychologist.

Mysticism was a fashionable topic during Victorian times. In some degree this was due to the revolutionary discoveries of the era, which meant that life for many had become uncertain. One by one the fabled beliefs of old, such as the assertion that blistering an invalid's skin with a red hot poker would drive away their illness, were challenged by scientific evidence to the contrary. An air of disquiet fell on the ardent believers of these once deep-rooted truisms, and many turned to spiritualism for solace. After all, given that it was now possible to achieve

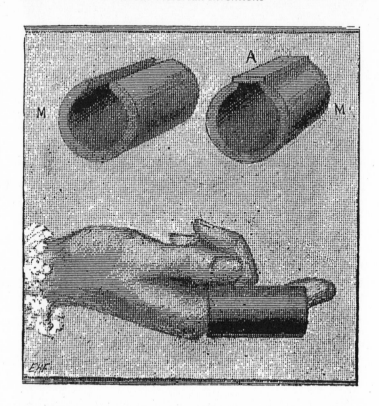

the hitherto impossible and speak directly with an absent friend through a very long cable, it was surely just a matter of time before modern technology provided a similar connection between the living and the dead. Victorian mediums claimed to have found the mythical link to 'the other side' by entering a hypnotic state and offering themselves as a channel for the spirits of the underworld to communicate through, just like a telephone. Naturally, Britain was fascinated by this bold claim, and the more impressionable individuals craved the opportunity to witness this spectacle for themselves.

Dr Ochorowicz's mesmeric device consisted of a hollow tubular magnet with a slit down the length, the edges of the slit being north and south poles respectively. To use the hypnoscope, the forefinger of the subject was placed through the tube of the magnet, so that both poles were united directly through the flesh of the finger. After two minutes the magnet was drawn off, and the person examined. According to Dr Ochorowicz, about 30 per cent of the test subjects examined in this way were found to have experienced some peculiar objective or subjective sensations during the time the hypnoscope was in place. Some experienced an itching or prickling of the finger, as if needles were perforating the skin. Others felt a sense of coldness, or of heat and dryness. Objective experiences were, on the whole, quite rare, but generally consisted of involuntary movements, complete insensitivity to touch, and paralysis of the finger. Dr Ochorowicz also observed that when a much larger magnet was placed under the feet of paralytics, their feet were warmed, even when sitting beside a roaring fire failed to have any such effect.

Overall, people who had been affected by the hypnoscope were generally considered to be capable of being put under a hypnotic trance by a hypnotist. This meant that they could, in theory, be alleviated of any number of deleterious disorders or mental grievances by the power of hypnotic suggestion.

The Gelatine Hygrometer (1887)

One common gripe that could not so easily be erased from the minds of the general public was the state of the weather, which, in certain parts of Northern Europe, tended to be permanently wet. In the mid-1880s an apparatus for registering the exact degree of moisture in the atmosphere was devised by the French meteorologist Monsieur Albert Nodon (1862–1934). It was known that gelatine could absorb a quantity of water proportional to the hydrometric state of the atmosphere, and in so doing the gelatine would increase in bulk and weight proportionately to that state. Gelatine was thus the chosen component of his weather meter, or 'hygrometer'.

To begin with, the gelatine was rendered durable by adding a small quantity of salicylic acid to it, and was then spread over a sheet of strong paper or cardboard. Spirals were cut from the paper, and enclosed within the apparatus as shown in the following image. There were four gelatine spirals, S S S S, each

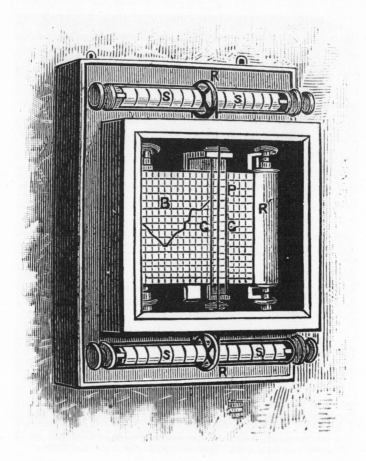

having one end secured, the other free and acting on pulleys, R R, between which a thread passed. The pulleys were in the same vertical line, and the thread carried a sliding stylus, P, which ran between parallel guides, G G. The stylus marked the position on the travelling band of graph paper, B, which was unrolled by clockwork from the roll, R. The twisting of the spirals produced by the amount of moisture in the surrounding air worked the pulleys, and raised or lowered the sliding stylus on the paper, thereby marking a continuous line, which represented the exact humidity of the atmosphere.

The Hydrophone (1887)

The gradual introduction of fresh water supplies to towns and cities played a major role in improving people's daily lives, as water towers, indoor plumbing, improved filtration systems, and domestic taps became standard across Europe and America. This early infrastructure, however, encountered a variety of teething problems, so eager inventors got to work on fixing these setbacks.

The figure represents an innovative contrivance devised by Herr A. Pares of Altona, Germany, which could detect leakages in water mains. This invention was known as a 'hydrophone'. A rod, A, made of sound-conducting wood, was held in a vertical position by a tripod, B. To its upper extremity was attached a metal box containing a microphone, M. The apparatus was completed by a rechargeable battery, E; a telephone receiver, P; a contact-maker, K, to close the circuit of the battery when an observation was to be made; and a box, D, to hold the parts. On moving the rod, A, over the water pipe to be tested, the sound of the leaking water caused the rod to vibrate. These sonorous vibrations were transmitted through the rod to the microphone, and so heard through the earpiece of the telephone when the contact-maker was closed and the electric circuit complete. The leakage was thus located and quickly repaired.

The Iceberg Detector (1885)

One place where it was essential never to encounter a leak was on board a ship. In the days before radar, satellites, and computer technology, seafaring vessels were often exposed to the perils posed by icebergs, specifically when sailing across the oceans of the northern hemisphere. The usual form of protection employed to guard against these deadly floating masses was quite simply the engagement of a night watchman who would turn on an electric light and keep his eyes peeled. However, in 1885, twenty-seven years before the *Titanic* embarked on her maiden voyage, iceberg detection experiments were carried out at Chesapeake Bay, off the coast of Virginia, with the aid of a gun and a large ear trumpet. The results of this curious investigation indicated that ships were good reflectors of sound, sending back audible echoes at a distance of up to a mile away. It was proposed to apply this method to identify the location of icebergs, which were, as was well known, particularly dangerous in thick weather. As the echoes of a steam-whistle or siren could be clearly reflected from an iceberg, thereby informing the ship's crew of its position, it was proposed that night watchmen should adopt this method during the voyage.

An alternative suggestion was put forward whereby the difference in temperature observed between the bow and stern of a vessel when approaching an iceberg might be caused to sound an electric alarm by means of an arrangement of thermometers or thermostats.

The Champagne-Bottle Drain (1887)

Across the Atlantic, the communicants of St Peter's Episcopal church in North Greenfield, Wisconsin, were experiencing some rather watery problems of their own. When the church was built in 1885 the community's funds were disappointingly low, and many economical shortcuts had to be made. As a result a cheap, inadequate drainage system was installed, and thus the congregation found that the church became unpleasantly damp. Furthermore, the building was predisposed to flooding during rainy months. The parishioners, upon drawing the minister's attention to the church's soggy condition, were promptly directed to go to his cellar, where, he informed them, they would stumble across a sizeable collection of champagne bottles, ample enough to construct a drain with.

'Our homemade champagne-bottle drain will last for many years,' the minister enthused, 'providing, that is, the drain is regularly cleaned and not allowed to get clogged up.'

A team of obliging volunteers were more than willing to assist in liberating the bottles of their contents and carefully removing the glass bottoms after each had been emptied. Once secured to the exterior of the church, the neck of each bottle was inserted into its neighbour's bottom, and so on, until an entire glass bottle drain had been constructed. The arrangement was such that the next time it rained, the water was permitted to flow from one hollow bottle into the next, thereby draining the church of all moisture. It was said that St Peter's has been an abstinent place of worship ever since, without a damp patch in sight.

The Mean-Time Sundial (1892)

When it wasn't raining, the sundial was a public convenience of great worth during Victorian times. Although not as widely used following the introduction of railway time in 1840, which brought about a national conformity to timekeeping, these astronomical clocks were still handily situated in municipal places, such as in the grounds of a church or outside a town hall. However, an ordinary sundial did not always keep the same time as a clock or a pocket watch did. This was due to the fact that the

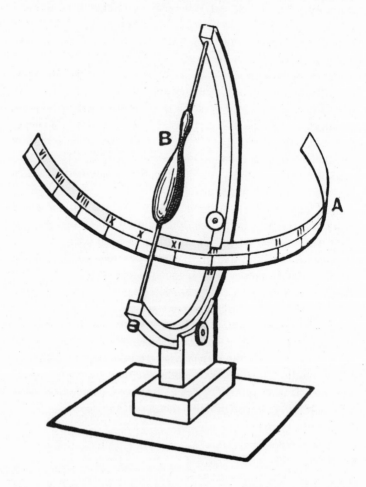

solar day varied with the position of the earth throughout the year. As a watch or clock did not rely on the sun to tell the time, it displayed the hours and minutes corresponding to the average length of the solar day, known as standard time. The 'equation of time', which was calculated by the time shown on the sundial minus standard time on a clock, gave the difference between the two, which could be anything up to sixteen minutes out.

By 1892 Major-General John Ryder Oliver (*c.* 1834–1909) had invented a sundial which made its own correction for the equation of time, and showed the mean-time like a clock. His apparatus is illustrated herewith, and consisted of a semi-circular arc, A, graduated with the hours, and a curved gnomon, B, that cast a shadow on the dial. Its curved form was such that the shadow of its edge compensated for the equation of time, whatever the position of the sun below or above the equator. It was adapted for any latitude, and by means of the clamping screws shown, the two arcs could be properly adjusted. Moreover, it could be set to show any required standard time, for example, that of Greenwich.

The Sun-Wound Clock (1884)

In Brussels, a new kind of mechanical clock was found to have been ticking for many months non-stop, without running down. This was because heat from the sun had been harnessed to automatically wind the timepiece, thus no battery or human intervention was required. To achieve this feat, an up-draught was obtained in a shaft behind the clockwork by exposing it to the sun. The air thus heated rose to the top of the shaft where it caused a fan to revolve, which automatically triggered a mechanism that wound up the weight of the clock until it too reached the top of the shaft. This then triggered a brake, which stopped the fan from moving until the weight had descended a little, at which point the fan was free to act again.

As long as the sun continued to shine, this spectacular sun-wound clock could be left to its own devices, keeping time for eternity.

Heat-Sensitive Paint (1887)

An inventor named Mr H. Crookes was successful in utilising the sun's heat to invent an unusual red paint, which grew darker as its temperature rose, until it reached 180° Fahrenheit, by which time it had turned completely brown. Mr Crookes' innovative product was ultimately adopted for use as a heat tell-tale, particularly within machinery. For example, if the bearings of an engine were coated with his new paint, the operator could tell if the engine was running cool or hot. A scale of tints for comparison, with the corresponding temperatures marked upon it, was a useful guide for using the paint, which regained its original red colour on cooling.

Mr Crookes later applied his new technology to indicate the strength of an electric current. It was well known that electricity caused the wire it flowed through to become hot, and the temperature produced increased with the strength of the current. Crookes theorised that if the wire was coated with his paint, the latter could be used to indicate the strength of the current in the wire. To test the theory, he coated a thin ribbon of copper with paint and passed a current of electricity through it. The paint was red to begin with, but soon become darker by degrees as the temperature of the ribbon rose with the strength of current flowing through it. When the ribbon had become totally black, this indicated that the current in the wire was too strong, proving a useful tool for electricians to work by.

Luminous Paint (1884, 1885 & 1887)

The Victorians knew that luminous paint was a substance that absorbed light by exposure during the day, and emitted it back again when it was dark. By the latter part of the century, this type of novel coating was being utilised for a number of domestic purposes, but few applications were thought to have been so resourceful as the one conceived of by the owner of a horse. This unnamed equine proprietor covered his mount's harness and bridle with paint, and the parts so treated were quite visible at night, thus the position of the horse was clearly indicated to the rider and other road users.

The Army also applied a dab of paint to the sights of rifles to enable them to be used in the dark more efficiently, while on the Philippine Islands, it was used to coat the interior walls of houses. In the event of an earthquake at night, the inhabitants were able to locate the doors, windows and other exits with ease, so as to escape in a timely manner.

Solar Panels (1884)

In 1884, at a meeting of the American Association for the Advancement of Science in Minneapolis, Professor Edward Sylvester Morse (1838–1925) of Salem, Massachusetts, exhibited a new device for heating one's home solely by the means of the sun's rays. The invention consisted of a large, flat surface of slate, similar in appearance to a modern-day solar panel, which was painted black and hung vertically on the outer wall of a building. The slates were inserted into grooves, like panes of glass in a window frame, and flues conveyed the heated air around the slates into each room of the house that required heating. During the winter, if the days were dry and fine, one solar slab measuring 8 feet by 3 feet in area could quite easily warm a room 20 feet by 10 feet, with heat to spare.

The Sun Motor (1884)

In New York, the inventor Captain John Ericsson (1803–1889) constructed an original 'sun motor' for converting the sun's heat into mechanical power. His machine consisted of a curved reflector, R, lined with window glass panes, silvered on the undersides like mirrors, and fastened to wooden staves that lined the curved iron framework. The reflector was supported on a central pivot, round which it revolved to follow the sun during the day. The glass panes reflected the sun's heat on to a cylindrical heater, H, mounted above the apparatus. This heater

contained a channel of compressed air, which, once heated, was transferred through a flexible pipe to power the steam engine, E, the average speed of which was found to be 120 revolutions per minute. It was anticipated that this solar engine might one day be used to supply a clean source of power to a whole host of everyday machines.

The Solar Battery (1884)

Though it was first observed as early as 1839 that it was possible to generate serviceable energy from the action of sunlight, it was a German electrician named Herr Sauer who, in around 1884, was said to have devised a voltaic battery that was charged up by the influence of the sun. The electrical power his battery generated was furnished by a chemical reaction caused by the rays of light, and thus Herr Sauer's battery was considered by his peers to be a significant advancement in the conversion of luminous rays into electricity.

The battery consisted of a glass vessel containing a solution of fifteen parts table salt to seven parts sulphate of copper in 106 parts water. Into this was placed a porous mercury cell. One electrode of the battery was made of platinum and the other of sulphide of silver, and both were connected to a galvanometer, which measured the amount of electricity produced. The platinum electrode was immersed in the mercury, and the silver in the salt solution. When placed in direct sunlight, the galvanometer needle could be clearly seen to deflect, and the brighter the sun, the stronger the current in the homemade battery was found to be.

Despite Herr Sauer's apparent success, it was the English chemist Edward Weston (1850–1936) who received a patent in 1888 for his invention of the world's first solar-powered cell.

Electricity from the Wind (1893)

Contrary to the widespread opinion of the day, a growing number of engineers began to doubt the future of an economy based entirely on fossil fuels, which were regarded by many as a seemingly endless source of power for the country's industrial plants. Some of these scrupulous engineers began harnessing a green source of energy from the sun, while others looked to the more traditional power of the wind, used for millennia to power the earliest windmills and sail mankind's most primal ships.

In 1887 the electrical engineer Professor James Blyth (1839–1906) began constructing the world's first wind turbine that would convert kinetic energy into electricity. He erected his pioneering windmill above his holiday cottage in the Aberdeenshire village of Marykirk. The turbine was used to drive a dynamo, which in turn charged an accumulator that provided electrical energy to power the electric lights in the cottage, free of charge.

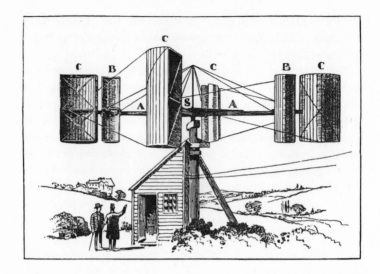

Professor Blyth's invention was completed by the 1890s, and comprised the ideal requisites for successfully generating electricity from the wind – namely, that his windmill could be left unattended to operate itself, and was situated on the Scottish planes, where the wildest gales blew.

In order to achieve the best possible motion, he incorporated on to his windmill the revolving cups of the Robinson anemometer, as shown in the figure, where the cups, C C C C, were semi-cylindrical boxes attached to four strong revolving arms, A A, each 26 feet long. The professor further augmented the power of the machine by adding a smaller auxiliary box, B B, to each arm behind the larger one. The round and vertical iron shaft, S, was attached at its lower end to a massive pit wheel. When the windmill revolved, it turned the shaft, which actuated a train of gearing that drove a flywheel 6 feet in diameter. From this flywheel the dynamo was driven by belting, which charged the accumulator and generated the required amount of electricity to power the house.

His success was such that Blyth offered his fellow villagers of Marykirk a surplus supply of electricity that would light the whole of the high street at night, but as the provincial residents believed electricity to be the work of the devil, the professor's offer was swiftly refused. He did, however, go on to receive a patent for his design in 1891, and was asked to construct a similar turbine above the lunatic asylum at Montrose, Scotland.

Professor Blyth's renewable energy plant was found to be particularly well adapted for lighting houses in the countryside and other regions where wind was more obtainable than coal or waterpower. Yet despite his revolutionary development, the first wind turbine to generate electricity for public consumption was not installed until 1951, sixty-four years after the conception of this original wind engine.

9

TECHNOLOGY AND COMMUNICATION

The Homacoustic Speaking Tube (1892)

The telephone remains one of the greatest and most useful inventions of the nineteenth century. Inspired by a seventeenth-century prototype, which literally comprised two tin cans and a piece of string, some of the greatest minds of the Victorian era conspired to develop a form of technology that would electrically transmit a human voice from one location to another. Though several men had vied to perfect this 'speaking telegraph', it was in 1876, after years in the making, that the Scottish inventor Professor Alexander Graham Bell was granted a patent for his variation of the telephone.

However, in the days before Bell's modern telephonic device had become widely available, the 'speaking tube' facilitated non-face-to-face communication in middle- and upper-class houses. When a member of the family wished to speak to, say, a domestic servant down in the kitchen, all they had to do was blow into their end of the long speaking tube, which would sound a whistle installed in the kitchen at the other end. This would swiftly summon the servant's attention, who would then be able to communicate with the caller, wherever they were in the house.

Though just about fit for purpose, the device was found to have certain shortcomings, such as the fact that there was only

one hole at each end of the tube, meaning that both speaking and listening was achieved through this single opening. So, by 1892 a new and improved kind of speaking tube was introduced. The illustrated 'homacoustic speaking tube' removed the need to blow into the tube in order to sound the whistle, W, at the other end of the line, as this now functioned automatically. One could also talk and listen without any change of position, owing to the employment of a new flexible earpiece, E, as well as a fixed mouthpiece, M. Moreover, this new design enabled the user to speak to occupants of any room in a house without requiring the installation of many mouthpieces and whistles, as was necessary with the old system; for now an automatic switch put the instrument in communication with any of the other tubes that ran throughout the building.

The Hands-Free Home Telephone (1893)

One of the initial objections to Professor Bell's telephone was that it was not a communal device, as only one person at a time could communicate with the individual at the other end of the

line. This prompted an engineer named Mr Graham, of Alfred Graham & Co., to invent a new telephone, which not only broadcasted loudly enough to be heard at the back of a large room, but was also to some extent hands-free, thus allowing both parties the freedom to take notes or read through important documents. This loud-speaking domestic telephone consisted of a microphone transmitter, T, and a magnetic receiver, R, with a trumpet mouth. In addition, two lines were employed, one for each correspondent, with a return wire common to both. By this arrangement the transmitter of correspondent A was directly in circuit with the receiver of correspondent B, and the transmitter of B with the receiver of A. Flexible conductors attached the transmitter to the line, and on pressing a button one only had to speak into its mouthpiece to attract at sufficient volume the attention of the correspondent, who replied by speaking into their transmitter in the same way.

Mr Graham also devised a switchboard, shown at S in the main illustration, by which a person could converse with any room or department of the building. The device was thus found to be especially useful in businesses, country residences, farms, and also on board ships. Indeed, the loud-speaking telephone was duly fitted on board the *Titanic*, and Mr Graham further proposed to install his appliance inside the helmets of deep-sea divers.

The Mechanical Telephone (1884)

Despite the fact that Professor Bell had been the first to reach the patent office, his fellow inventors were undeterred, and they continued to work on and improve Bell's initial blueprint. For instance, the mechanical telephone, which was invented in around 1884, was said to be capable of transmitting speech over a considerable distance along a wire without the aid of electricity. It consisted of a mouthpiece, A, which had a central aperture allowing the sound waves to pass to the diaphragm, D, which was secured round the edges to the mouthpiece. The diaphragm was about 7 inches in diameter, and was made of spruce pinewood, which was both strong and sonorous. The mouthpiece and diaphragm were carried by a bed piece, B, which was fixed to the wall. The bed piece was recessed at both sides, F G, and contained a central aperture for the passage of threads connecting the line wire to the diaphragm. The front recess, F, afforded a space between the diaphragm and the centre of the bed piece for free action of the diaphragm, giving clearness of articulation, and the after recess, G, allowed for the telephone being in contact with the wall by just a narrow space, thus preventing external vibration.

The line wire, which carried the vibrations of the voice to the receiver at the other end, was made of copper strands twisted

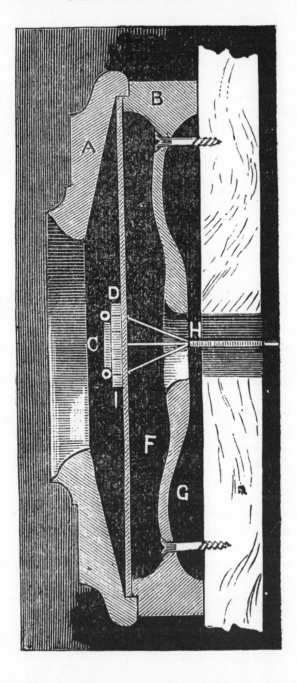

together and coated with varnish. This was connected to the diaphragm by silk cords which diverged from the wire, as shown at H. These strands were secured to a metal ring, C, between which and the diaphragm was interposed a leather stop, I.

Through the receiver at the other end of the line, the correspondent could perfectly hear what the speaker had said, and could respond accordingly through their own mouthpiece. The invention was fundamentally an improved version of the 'string telephone', and hailed from New York.

The Osmium Telephone (1884)

An American named Mr George Lee Anders presented another variant of Bell's telephone. This gentleman had discovered that the rare and expensive metal osmium made for an excellent

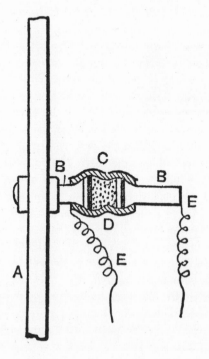

telephone transmitter when broken into grains and placed in a box between two aluminium or platinum electrodes, and attached to a diaphragm. The figure illustrates this arrangement in cross section, where A depicts the diaphragm, shown endwise; B B the two metal electrodes; and C D an ebonite or wooden box, containing the granulated osmium. The wires, E E, conveyed the current through the osmium by means of the electrodes, B B.

When the speaker's voice vibrated the diaphragm, the loose metal grains in the box were correspondingly agitated, and acted as a microphone on the electric current, modifying it in sympathy with the waves of sound. This current was sent through the primary circuit of an induction coil, and the secondary circuit of the same coil was connected in circuit with the telegraph line. The vocal currents passed along the latter to the receiving telephone, and made themselves heard as speech.

Remarkable Uses of the Telephone (1884)

Towards the end of the century telephony had made sizeable progress, and offered copious – not to mention eccentric – possibilities in the world of communication. With the help of the telephone, an entire match was played between the Cardiff & County and the Swansea Chess Clubs in 1884. The great novelty of this arrangement was the fact that the match could be played instantly, despite the opponents not even being present in the same room, or indeed the same town. The postal authorities had agreed to place the two municipalities in telephonic communication, and the match resulted in a victory for Cardiff. This historic event became the first chess match in the world to be played over the telephone.

While the British were taken by the originality of such an occurrence, once again it was the Americans who were pushing technological boundaries further than any other nationality. By the 1880s entire marriage ceremonies conducted over the telephone were suddenly no longer a novelty in the States. Europe, too, began using the telephone for more radical purposes. The King and Queen of Portugal, being prevented by a death in the family from publicly attending a musical performance at a local theatre, had the music wired by telephone to the palace, where they listened to the entire concert in privacy.

The Automatic Telephone Exchange (1892)

From the earliest days of the telephone right up until the mid-twentieth century, switchboard operators were the first point of contact for a subscriber to a telephone exchange. When the caller wished to speak to another subscriber, they were put through to the operator at the exchange to explain their requirements, and the operator would connect the caller's telephone with that of the other subscriber. In 1888, however, an American inventor named Mr Almon Brown Strowger (1839–1902) boldly suggested that it might be possible to dispense with the assistance of operators in telephone exchanges altogether by introducing an automatic electric mechanism to take their place. Four years later the world's first example of this revolutionary technology was set up in Indiana. There were a mere forty-five subscribers at the time.

With the new so-called Strowger apparatus, the subscriber could make the desired connection themselves, without the need of a human operator. There was one line installed for each subscriber, and so long as these were kept in good working order, it was said that no staff members were required to operate the exchange.

Each subscriber was provided with five keys or contact makers for opening and closing electric circuits. Four of these were marked 'units', 'tens', 'hundreds' and 'thousands'. Supposing the number of the subscriber the caller wished to speak with was 131, the caller was required to press the unit key once, then tens key thrice, and the hundreds key once. In so doing, corresponding electric currents were transmitted to the exchange, which actuated a small automatic apparatus that connected the caller's line with that of number 131. When the telephone conversation was over, the subscriber, by pressing the fifth key, broke the connection and restored the original condition of the circuits.

By 1897 Mr Strowger had introduced his automatic system

to Britain, and fourteen years later the Automatic Telephone Manufacturing Company Ltd was supplying automatic telephone equipment to consumers across the country, hailing, in a small way, the beginning of the end of the telephone operator.

Wireless Telegraphy (1893)

When the telephone was first introduced it was found to be extremely receptive to the currents induced in neighbouring lines, causing a disagreeable pattering noise which often threatened to drown out the spoken message. Moreover, the words spoken on another line could often be overheard, and this cross-talk was most unwelcome by those not partial to morsels of juicy gossip. However, when properly investigated this strange phenomenon was found to offer a great deal of promise to those inventors and engineers who were working on wireless technology.

Thomas Edison was one such gentleman who began conducting experiments based upon this irritating electric induction through the ether from one wire to another, and successfully devised a system of telegraphing messages through the air to and from a moving train. This became known as 'grasshopper telegraphy', as the signals were made to 'jump' between the lines.

A fellow inventor also proposed to attempt to send a wireless telegraph across the entire breadth of the Atlantic by means of two circuits, one on the American, the other on the European seaboard, but this was an ambitious project at the time, and the results were unsuccessful.

By 1893 Mr William Henry Preece F.R.S. (1834–1913), engineer-in-chief to the Post Office, succeeded in sending a wireless telegraph 3 miles through the air between Lavernock Point, near Cardiff, and the island of Far Holme. A telegraph circuit was run along the shore at both places, the one on the mainland being supplied with strong signal currents from a dynamo. The message sent along the wire was received by induction on a 'sounder' telegraph instrument connected to the island's line. The sending of the telegraph was hailed as wireless, as it was completely independent of any cable across the water.

Cynics of the era asserted that luminous semaphore signals from an electric or oil lamp could be read just as easily and at a greater distance by night, but supporters of the new technology argued that a wireless telegraphy system would also be feasible during the day.

The Travelling Telegraph (1885)

In the 1880s Edison's ingenious grasshopper system for keeping up telegraphic communication with moving trains, throughout the whole length of a railway, was introduced on the New York and New Haven Railroad. It consisted in running a telegraph conductor, C C, in a closed trough between the rails of the line, and attaching a long parallel coil of wire to the bottom of the telegraph office carriage, which formed part of the train. This parallel coil was fixed so as to move closely over the conductor as the train travelled along. As it did so, telegraphic signal currents were sent into the conductor from the railway stations that wished to send a message to the travelling train, which induced corresponding currents in the parallel coil that hung from the bottom of the travelling telegraph carriage. This was in circuit with a delicate relay within the carriage, which triggered a local battery on board the train. The currents of the battery actuated an ordinary telegraph instrument, and thus the message was successfully delivered to the train's attendant clerk. In this way,

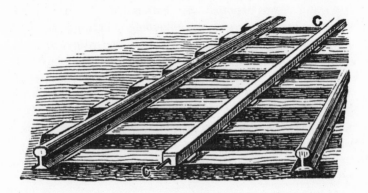

a person travelling by rail could maintain telegraphic communication with the stationary world, and also receive the latest news and prices of stocks. By a reversal of the process, the clerk could telegraph back from the moving train.

Sea Telephony (1884, 1885 & 1887)

The prospect of sending telegraphs not just through the air, but also through water, had excited considerable attention. However, in 1884, at the British Association Meeting in Montreal, the English physicist Lord Rayleigh (1842–1919) expressed his opinion that it was impossible to telephone speech through more than 40 miles of a sub-Atlantic cable; a theory that was soon dismissed when experiments between Dublin and Holyhead demonstrated that speech could in fact be transmitted 60 miles on a line of the same type.

The distinguished Professor Bell became engaged in his own maritime experiments. With the help of an assistant, he rowed two boats out into deep water and anchored them just over a mile apart. In one boat was Bell and a telephone, and in the other his assistant with a voltaic battery. One terminal of Bell's telephone was connected to the water, over the bow of his boat, and the other terminal entered the water at the stern. Whenever the battery in the assistant's boat sent an electrical charge into the water, Professor Bell reported that he could hear a musical note in the telephone, which was held to his ear.

In 1887 it was reported that Edison had also been busy perfecting his own system of telephoning underwater, so as to enable ships to communicate with one another wirelessly, even in stormy weather. He had discovered that submarine telephonic signals could be produced by means of small explosions, set to mimic signals of the Morse telegraph code. The impulses from the explosions were transmitted through the water and picked up by a telephonic apparatus on board the ship that was to receive the coded message.

The Scoto-Irish Telephone (1893)

By 1893 telephonic communication had developed to such an extent that Ireland and Scotland were at last able to converse by telephone, as the Post Office, which at that time licenced the national telephone networks, had submerged a speaking cable between Port Kail and Donaghadee. These two landing places were connected, by overland lines, to Belfast on the Irish side and Glasgow on the Scottish, and the spoken voice, travelling from one country to the other, was said to be heard very satisfactorily.

Telegraphing 7,000 Miles (1884)

In the days following the Industrial Revolution, merchants across the globe were demanding faster, more reliable modes of transport to convey their produce from one part of the world to another. In order to stay ahead of their competitors, they also required a means of speedy communication between their customers and their business partners. Hence by the middle of the century it was proposed to construct a massive telegraph line that linked England directly with India. By 1884 this great feat in telegraphy had been accomplished in the form of an Indo-European line which ran overland for 7,000 miles. London, Emden on the Baltic, Odessa on the Black Sea, Teheran in Persia, Karachi, and Calcutta were all connected, and messages were sent and received over this enormous length of wire at the rate of between twelve and fourteen words a minute, using Morse code. That such a marvel of engineering was realised was testament to how far technology had come in a relatively short space of time, for in 1866 only one message could traverse the same distance in twenty-four hours.

Prescribing by Cable (1887)

The Atlantic Cable was another great undersea telegraph line that had been laid at the same time as the one aforementioned, connecting Europe with North America, and by 1887 it had found a novel use, one that greatly beguiled the Victorians. A lady from New York, while travelling on the Continent, was taken ill of fever in a German town. Her accompanying relatives were greatly distrustful of the German medical attendance, so they sent instant telegraphs via the Atlantic Cable every day to their dependable family physician in New York, and received in reply full particulars of the treatment to be pursued. This communication occurred almost instantly, whereas in the days before the cable had been installed it used to take over a week for a message to be conveyed one way by ship. Thanks to the physician's timely advice the patient eventually recovered, and returned to America in full health.

As trusted family doctors were believed likely to know a patient's constitution better than strangers, this new use of the telegraph was greatly publicised at the time, in the hope that it would inspire others to follow the example of the New York family.

The Telautograph (1887)

Professor Elisha Gray (1835–1901), a well-known American electrical engineer, invented a remarkable automatic telegraph machine that was dubbed the 'telautograph'. Its distinguishing feature was that it enabled a person to write a message in his own hand on to the transmitter, and on the receiver, situated at a distant place, an exact facsimile of the sender's handwriting was produced simultaneously by means of an automatic stylographic pen. This second pen would, by means of an electrical signal, automatically follow the movements of the first, thus reproducing the outline.

Though this was an amazing achievement at the time, it was not an original concept. In fact, several apparatuses had previously been devised for this purpose, including the telegraphic pen of Mr Edward Alfred Cowper (c. 1820–1893), which was introduced in around 1879. Professor Gray's apparatus brought significant improvements to the design, though the specifics of this early fax machine were kept a closely guarded secret until 1893, when the telautograph was first demonstrated to the public in Chicago at the World's Columbian Exposition.

Following the success of his invention, the professor founded the Gray National Telautograph Company, and over the years his machine proved to be useful for transmitting instant signatures from one office to another. Institutions such as banks, hospitals and Army barracks put the telautograph to regular use, allowing bankers, doctors and military officials to sign vital documents in the blink of an eye. The business operated under the name The Telautograph Corporation until it was taken over by Xerox in the 1990s.

The Picture Telegraph (1892)

The next stage in the development of Victorian telephony was to attempt to discover a way in which to combine the two greatest inventions of the nineteenth century – the camera and the telephone. The 'electroartograph', as it was termed, was a machine designed in about 1892, which was said to possess the ability to transmit a photograph down a telegraph wire by means of electricity. This remarkable telegraph engraving machine was the work of an inventor named Mr N. S. Amstutz, from Cleveland, Ohio, and was relatively successful in its objective, as the figures show.

The process consisted of photographing the picture to be transmitted and transferring the image on to a specially prepared film of gelatine. When the film was exposed to light, certain shades on the gelatine were hardened. After being washed with water to remove the portions not affected by the light, the film displayed the picture in relief.

The next step was to vary the strength of the electrical current in the telegraph wire according to the variations of light and shade in the picture, or in other words, according to the heights and depths of the etched film. To achieve this, the point of a stylus was passed over every part of the film, which tripped up and down in replication of the degree of relief. The stylus's

movements sent electric currents down the line, the strength of which depended on the variations in height on the relief. At the receiving end of the electroartograph the current passed through an electromagnet, which pressed with more or less force on a travelling engraver that etched an exact copy of the design on to a surface of wax, from which a print could be made.

The image on the left represents a portrait on the gelatine film, which was transmitted through a 20-mile line. The reproduction is shown on the same scale on the right.

Telephotography (1893)

In 1893 an American electrician named Professor William White Jacques (1855–1932) delivered an address before the German

Technical Society of Boston in which he referred to experiments he had seen on telephotography, or the transmission of photographs to a distance by means of electricity.

He spoke of the photographic laboratory he was taken to, which consisted of two rooms. In one was a camera, a small developing closet, and on a table in the middle of the room was a cubical box, in one side of which was a slit the size of a postcard. From this box two wires stretched across the room to a wall. The wires passed through the wall and terminated within a similar box standing on a table in the middle of the adjoining room.

Professor Jacques was handed an ordinary postcard and asked to write a short note upon it. After penning 'Good morning, how do you do?' the experimenter then placed the card in front of the camera, where it was well illuminated by an electric lamp. He took a photograph and promptly departed into the developing closet, before returning to the room with a negative. This he dropped into the slit in the box. Professor Jacques was then directed into the adjoining room, where, issuing from the corresponding box on the table in the middle of the room, was a piece of thin paper the size of a postcard, on which appeared in facsimile the words 'Good morning, how do you do?'

The Stenotelegraph (1884)

Using the traditional system of telegraphing, approximately fifteen individual signals were required to send the average English word, as the average letter required three signals, and the average word consisted of five letters. In an attempt to simplify and thereby quicken the process, Sig. Michela of Ivrée, Italy, set himself the task of inventing a revolutionary telegraphic shorthand. He proposed to simultaneously telegraph, by means of a keyboard similar to a piano of twenty keys, any speech in any language as fast as it could be spoken. The apparatus he devised was said to be capable of sending a remarkable 10,000 words per hour, more than double the speed of a modern-day professional typist.

The principle of Sig. Michela's 'stenotelegraph' was to electronically transmit the phonetic sounds common to every language by means of signals, not letters. The operator heard the phonetic sounds in every word, and telegraphed each by pressing down the corresponding keys. At the receiving station short horizontal lines on a moving strip of paper represented each phonetic sound, which, when put together, could be translated back into the original language.

The Acme Thunderer Whistle (1892)

Despite the ever-developing technology of this great scientific age, there were some communicatory inventions that did not rely on wires or electricity to achieve their aims, and were as simple as they were effective. One that stood the test of time was a very loud silver whistle, which could sound a thunderous blast over a mile in distance with only a slight expenditure of force.

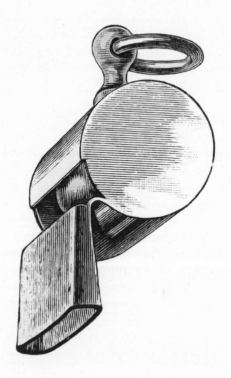

The new 'Acme Thunderer', as depicted, was invented in about 1892 with the intention that it should prove useful to cyclists, firemen, and sports referees. It was also adopted by the police force in order to replace the much quieter whistles and the policeman's cumbersome rattle, and later, in 1912, by the officers and crew of the *Titanic*. It was said that the Thunderer was used by the first mate of the *White Star Line* steamship to summon the help of passing vessels. Owing to the intensity and clarity of its blast, the whistle saved the lives of 700 people on that ill-fated night. It was recently sold at auction in London for £3,100, making it one of the most expensive whistles in history, and thus earning it a well-deserved place among the most remarkable of Victorian inventions.

10

TRAVEL AND TRANSPORT

The Tricycle Chair (1885)

The early nineteenth century had given rise to the steam locomotive railway that crisscrossed the Western world, linking villages with towns, towns with cities, and had even made navigation between countries quicker and easier than ever before. Provincial townsfolk who had never ventured beyond the end of the road that led from the factory to their door could now step aboard a train and head to any manner of destination they

pleased. Unique to London was the exceptional Underground network, which, from 1863, had made the sprawling capital simple to navigate. These were just a few new constructions that helped to revolutionise the way people travelled, but by the end of the century even more courageous flights of fancy were craved, and a few people dared to look to the skies.

Yet it wasn't just the radical new modes of transportation that captured the imagination, for it was often the smallest improvements to everyday life that made the biggest difference. The 'Bath chair', for instance, was a time-honoured eighteenth-century vehicle that was invented specifically for invalids, similar to a modern-day wheelchair. While it served its purpose in transporting those of limited mobility from one place to another, it was described as being a very 'slow coach', and some convalescents expressed a desire for a swifter transit. This prompted the development of the speedier 'tricycle chair', as illustrated, which could be driven quite easily at the rate of 5 miles an hour. The tyres were made of India rubber, and the chair was mounted upon tricycle springs, hence the motion was said to be easy and pleasant for the rider. The invalid was seated in a wicker chair in front of the driver, who worked the pedals and steered the machine from the back.

At a trial in Warwickshire, a workman succeeded in propelling this new vehicle, containing a sitter weighing nearly 12 stone, all the way from Coventry to Birmingham and back – a distance of 35 miles – at the rate of 8 to 9 miles per hour.

The Quadrant Tricycle (1884)

Another novel Victorian roadster was termed the 'quadrant tricycle'. It had been introduced to the streets of Britain by 1884, and consisted of three equally sized wheels, so that the weight of the commuter was divided between them. This gave the third wheel a better bite of the ground, rendering the act of steering light and easy.

The Bicycle Railway (1892)

In the heart of New Jersey, between Mount Holly and Smithfield, there once existed a railway for cyclists. Inspired by the success of the railroad, an enthused American inventor came up with this original way to travel, which united the simplicity of the bicycle with the intricacies of the locomotive. His single rail was mounted on a wooden fence, and the bicycle was inverted in the manner shown. Each passenger was, in effect, his own locomotive, and thus combined healthy exercise with travel. There was no danger of falling off and no need to bother with any form of steering, thus commuters could freely look about and enjoy the air.

Though plans were underway to double the track for coming and going passengers, and also to provide stations at intervals which offered fresh supplies of vehicles, the bicycle railway closed

down just five years after its opening. Similar trials were made in England across Norfolk, Great Yarmouth and Blackpool, but by the turn of the century this unusual mode of transportation fell out of favour and the tracks were ultimately dismantled.

The Road and River Cycle (1893)

This was not the last the world would see of combination vehicles, for the conjoined boat and tricycle, shown overleaf, was

introduced in America for travelling across roads and rivers in a single vehicle, thereby eliminating the inconvenience of having to travel for miles around stretches of water.

The principle of the contraption was simple: two boats were fastened to either side of a tricycle, the wheels having been fitted with paddles so that when the vehicle was in the water the paddles could be guided by the steering wheel, propelling the vehicle smoothly across the waves. When on the road, the vehicle operated as an ordinary tricycle, with the boats serving to hold luggage, fishing tackle, picnics, and other paraphernalia, making this leisurely pleasure cruiser perfect for summer jaunts.

The Unicycle Sleigh (1884)

For the long winter months, an American named Mr J. B. Bray invented a more hard-wearing combination vehicle in the form of a sleigh crossed with a unicycle. His adapted velocipede promised to become most useful in America and Canada, where long frosts were regularly experienced during the wintertime, making journeys difficult for commuters.

The unicycle sleigh consisted of a bicycle frame and saddle, supported by four sleigh runners and a central wheel, studded round its rim with spikes that bit into the ice. This was the driving wheel, which was operated by foot pedals in the ordinary way. The two front skates, or runners, were made to slew round at the will of the driver in the act of steering.

Treadle Driving (1885)

One of the most popular and convenient forms of road travel in the nineteenth century was the horse and cart. The Hansom cab was a common sight on the streets of Britain from 1834 until the first decades of the twentieth century, and during that time many amendments were made to this customary carriage.

Inspired by the pedals on bicycles, a new method of driving horses by means of the feet rather than the hands was introduced. The illustration shows the visionary treadle-driving arrangement

in action, which was brought out in 1885, a year before the German inventor, Herr Karl Benz (1844–1929), unveiled the first modern motor car. The driver's feet rested on a firm board, and the horse was guided by raising or lowering the toes, thus bearing on one or the other rein by means of straps in connection with them, which passed over the pulley mounted on the front board of the vehicle, as shown. The driver's hands were then quite free, and could be inserted into the pocket of his greatcoat to keep them warm in cold, stormy weather. Furthermore, when the driver left the vehicle, an automatic gentle bearing on the horse's mouth served to keep the animal still and quiet while unattended.

The apparatus, it was stated, could be attached to any vehicle in just a few minutes, and was so simple to use that a person of any age and ability could operate it with safety.

The Hippometer (1893)

A French cavalry officer named Captain Buisson, of the 5th Chasseurs, invented another useful accessory for one's horse and cart. His 'hippometer', as it was called, was a pedometer for horses, which was used not only for measuring the distance

traversed, but also the number of paces the animal had taken since the start of the journey. The hippometer resembled a watch, as seen opposite and was similar to the ordinary pedometer, which indicated the distance walked by a person. The mechanism was actuated by the shock of the footfalls, and was carried in a leather case, which was strapped to the horse after the manner shown above.

Spectacles for Horses (1893)

Owing to its incredible strength and stamina, the horse remained a creature of great value during Victorian times, and as a working animal its uses were plenty. The health and general wellbeing of this most serviceable animal was therefore of the utmost importance, and it was reported in 1893 that the concerned owner of a short-sighted horse went into an optician's store and ordered a pair of equine spectacles.

This prompted an extended study of the eyesight of horses, and Mr Dollond, of a well-known firm of opticians, became convinced

that horses, along with other animals, suffered like humans from short and defective sight. Furthermore, the optician was of opinion that degenerative vision was often the cause of a horse shying or bolting from a frightening-looking object. It was a fact that when a nervous horse was led gently towards whatever it was that had spooked him, and given an opportunity of seeing it clearly, his fear would quite disappear. Mr Dollond appreciated that this was not always practicable or convenient, so he devised the bi-focal spectacles shown in the photograph, which completely covered the eyes of the horse, in order that he might see distant objects more clearly.

At first the horse was said to have been alarmed by this strange new addition to his harness, but once he had got used to the spectacles, and the thick straps that held them in place, it was alleged that he strongly objected to *not* wearing them.

Mr Dolland's bi-focal equine lenses made the road appear to rise up in front of the animal, as if walking up a slope, and by wearing these glasses the horse was thus induced to step up high. It was not long before horse trainers and Hansom cab proprietors adopted the use of these spectacles for the promotion of high-stepping – a desirous trait that was deemed aesthetically pleasing – when drawing carriages through the streets.

The Cab Communicator (1887)

For added convenience to the Victorian commuter, an electrical device for enabling a fare to communicate with the driver of his cab was fitted inside a number of horse-drawn vehicles. This invention consisted of an indicator, upon which were written the words 'Stop', 'Right', 'Left', 'Turn Round', and other appropriate commands. On pressing the corresponding button, a small bell tinkled and an arrow flew round and pointed to the required word, which was printed close to the driver's eye line.

The Camel-Drawn Monorail (1884)

A French engineer named Monsieur Charles Lartigue (1834–1907) devised an original method of transporting goods across wild countries where long stretches of desert meant it was impractical to employ the use of a horse. His invention was a camel-drawn vehicle that ran for miles on a single rail. It was originally intended to transport esparto grass, commonly used to make paper and baskets, from the Algerian hills all the way to the coastal towns, but the design was also applicable to any open country that was free of fences.

The rail was made of iron, and each section was supported on two iron legs, which could be easily stuck into the ground by a workman. As such, the rail ran 31 inches above the ground, and each length of rail weighed 33 lb, while the supports weighed around 31 lb. The railway was cheaply built, as no preparation of the ground was necessary, and six men could lay 2½ miles of line per day.

Once the rail had been completed, a pair of iron panniers, fixed on a double steel bar, was hung over the rail, and ran on two wheels. A single camel could draw sixty of these panniers loaded with grass for 56 miles across the desert. The centre of gravity of the load was below the rail, and there was thus no danger of capsize.

The Cable Tramway (1884)

A comparable monorail was built in Europe, though this one was powered by an electric motor. An experimental line was exhibited at the Palais de l'Industrie, Paris, as part of the International Agricultural Exhibition, where it was even shown drawing commuters. The passengers sat back-to-back in iron chairs, or panniers, thus helping to equilibrate the load upon the rail. The whole load, including freight and cars, was supported on the single rail by small wheels, which ran in a groove along the surface of the rail. A train of panniers was drawn by a dynamo-electric locomotive, which contained, besides the motor, the starting and stopping switches, the brakes, and a rheostat for graduating the strength of current, and thus the speed of travel. At full throttle the monorail was said to be capable of reaching speeds of about 7 miles per hour.

It was hoped that such a railroad would one day prove useful for transporting commuters around busy cities, thus easing traffic congestion on the streets below, as well as through mines, forests or prairie lands.

The Telpher Line (1884)

At Weston, near Hitchin, Hertfordshire, was erected a trial line of the aerial 'telpherage' system, which was invented by the Scottish engineer, Professor Henry Charles Fleeming Jenkin (1833–1885). This novel mode of airborne transportation, which ultimately evolved into the modern cable car, was designed to carry goods or passengers across vast distances, and the original line is illustrated opposite.

With the use of tall posts, Professor Jenkin's aerial tramway was raised high over the roofs of the buildings, allowing trucks

full of grain, coal, or food to be transported over vast distances from one place to another. The wire ropes, which served as rails for the trucks, conveyed the electric current to the electric motor at the front of the ten-truck locomotive, and a stationary dynamo at one end of the line supplied the current. The train could reach speeds of up to 4 miles an hour.

The trucks terminated at a railway station, where a team of labourers would tip the contents into waiting waggons, for the goods to continue their journey by land.

The first operational telpher line was constructed in the village of Glynde, East Sussex, in 1885, and it was highly anticipated that telpherage would play an important part in the future of locomotion.

Pike's Peak Railway (1892)

Since the year 1830, when the first steam passenger service on the Liverpool & Manchester Railway line began operating, Britain had witnessed an increasing public desire for taking pleasure trips. Mr George Bradshaw's famous railway handbook offered would-be passengers a tantalising glimpse at the exciting new places across the whole of the country that were just waiting to be explored.

Traditionally, the only ways for travellers to reach the summits of mountains and other remote places of natural beauty were either by foot, on horseback, or, if they were lucky, aboard a sedan chair. In 1880, however, a funicular railway opened on Mount Vesuvius, which transported thousands of tourists to the summit of this infamous Italian volcano until the service ceased in 1944.

Similar modes of transportation also carried nineteenth-century visitors up the Swiss mountains Rigi and Pilatus, and by 1892 a cogwheel railway was opened in Colorado, which ran from the spa city of Manitou Springs to the summit of Pike's Peak Mountain. This snow-covered prairie landmark was calculated to be 14,336 feet above sea level, affording a magnificent prospect of the Rocky Mountain chain on the west, and the illimitable plains on the east. Scenic horse trails were soon made along its streams. The picturesque route known as Ruxton, bordering the beautiful Ruxton Creek, became the favourite path to the summit, some 7 miles distant, and it was by this route that the new railway was built.

The track was made from steel, and the locomotive climbed the steep grades by means of six cogwheels working into two serrated rails laid between the ordinary rails, after the manner of a rack and pinion. The line was almost 9 miles long, and rose 7,500 feet up the peak, with the steepest grade having a rise of 1 foot in every 4 feet of horizontal distance.

There were two coaches to a train, each holding fifty passengers, and the journey was made in less than two hours, including a stopping point at a halfway house. Splendid views were to be had on the way up, and eventually a hotel was built on the summit for the accommodation of tourists. There was little doubt on the part of the railway engineers that Pike's Peak would henceforth become a popular destination for both the ubiquitous tourist and the seeker of medicinal mineral waters.

The Ghost Train (1892)

Passenger trains of the early nineteenth century were generally very uncomfortable to ride in. The seats were wooden and exposed to the elements, and the motion was jolty. In later years first-class passengers were offered a more comfortable trip when upholstered seats, armrests and oil lamps were fitted inside the carriages, and railway companies soon began competing with one another to offer commuters the most pleasing ride.

In an effort to attract only the wealthiest of passengers, an American rail company set up a new kind of luxury passenger train that ran between New York and Boston from 1891. This new 'ghost train', so-called because of its spectral appearance, operated every afternoon, except Sunday. The creamy-white carriages were decorated with gold lettering and other embellishments, and staff uniforms were entirely white. The exteriors of the superior 'Pullman' carriages were brightly illuminated after dark by gas jets, and the wheels were said to have been made from paper pulp. During the hours of darkness, and especially when the air was filled with mist, the vivid glow of the gas lamps inside the carriages, coupled with the soft illumination from the white exteriors, made the train appear distinctly ethereal to anybody witnessing its passing.

The train, which was warmed with special heaters, consisted of a sumptuous combination car, with a luggage van and smoking room. The latter was carpeted, furnished with willow chairs, and upholstered to resemble the finest hotel room down to the last detail. The drawing-room cars came with many conveniences such as revolving and reclining chairs, and the passengers were waited on hand and foot.

Despite the commensurate luxuries of first-class travel, the trains were marketed as having no distinction of class; the humblest traveller was free to enjoy all the comforts the train had to offer ... providing they could afford a ticket.

Sadly, this spectral mode of opulent travel was not destined to endure. The service ceased in 1895 due to the enormous costs involved in keeping the pure white carriages spotlessly clean.

Heating Railway Cars by Gas (1887)

Most passenger trains of the Victorian era were not equipped with a central heating system, and thus travellers were often chilly during their journeys. That was until Mr William Foulis (1838–1903), a wealthy gas engineer from St Andrews, Scotland, devised a new system for warming hitherto unheated railway cars using the waste heat from the gas lamps that hung from the carriage ceilings. A water boiler was placed over the gas flame, and an arrangement of pipes circulated the heated water from the boiler, through the carriage and underneath the seats. It was found that the ordinary size of a gas flame was quite sufficient to heat an entire compartment, and mixing a little glycerine with the water prevented it from congealing when the carriage was not in use.

This early gas central-heating system was effectively trialled during the winter of 1887 aboard the trains of the Glasgow & South Western Railway. The temperature of the carriages was successfully kept from 52° to 60° Fahrenheit, even during very cold weather.

The Foot-Warmer for Trains (1885)

The accompanying figure is a sketch of a personal electrical foot-warmer that was developed for the use of Victorian passengers aboard nippy trains. The device was to be placed in the bottom of a railway carriage in order that the passenger's feet could be rested comfortably upon it. The inventor hoped that by the time electricity was supplied to railway carriages for lighting purposes, his device, which he originally intended to be a bed-warmer, might prove serviceable to commuters.

The foot-warmer consisted of a large metal box or hot-air chamber, perforated with air holes at the top and bottom, *g g* and *f f*, the latter being inlet and the former outlet holes. In the middle of this box was a smaller chamber, consisting of a block of fire clay that was also perforated with holes, *i i*. These holes contained coils of bare wire connected to the outside terminals, *t t*, by which the electric current was sent through them. The

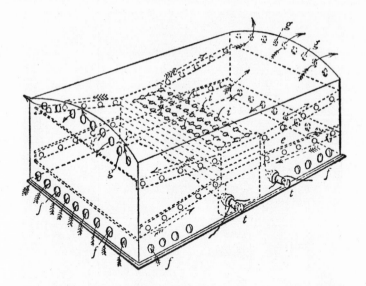

current heated the coils of wire, and thus the air circulating around the box in the direction of the arrows was warmed up, and escaped by the outlet holes *g g*.

This device ultimately proved a great success, and from the late nineteenth century until the 1920s electrical foot-warmers played a key role in railway comfort. Passengers were given the opportunity of hiring these boxes prior to boarding, although it was said that a porter had to shake the early models vigorously before they sprang into life.

The Travel Holdall (1893)

The following illustrations show a new type of travel convenience in the form of a 'holdall', which could be purchased by rail passengers before they made their journey, and was particularly useful for those who could not afford to travel in the comfort of first class. Once they had taken their seat, the holdall could be converted into a reclining leg-rest, as shown. It was invented by Dr Batten, who also claimed the holdall could also be transformed into a luncheon table, or, in cases of accidents, as a splint for a broken leg.

100 Miles an Hour by Rail (1892)

In the very first days of rail travel it was often presumed that misfortune was likely to befall a railway commuter, for early Victorian locomotives were speedy vehicles that whizzed across the country at breakneck speeds of around 30 miles per hour. This was a major increase in pace in comparison to the more traditional rides such as the stagecoach and the steam-powered tram. Indeed, some of the first railway travellers became distressed when they learned of the terrific velocities at which they were to be catapulted across the land. Such reservations dwindled, however, as passengers became accustomed to rail travel, and speeds of up to 60 or 70 miles an hour became standard on most major routes. By the end of the century it was said that the dream of reaching that elusive speed of 100 miles an hour could finally become a reality.

The Americans foresaw an electric railway that could be constructed between Chicago and St Louis, a distance of almost 300 miles, on which engines could be made to travel at this hitherto unreachable speed, and would thus complete the journey in less than three hours. The cigar-shaped cars were to be propelled by electric motors connected to the axle of two driving wheels on each, and the current was to be supplied from a central station at a convenient point on the line. Telephones and electric lights were, of course, to be provided, and the surplus electric power would be supplied to the district surrounding the central station.

Despite America's high-speed vision, it was a British steam train that became the first rail vehicle to officially travel at 100 miles an hour. On 30 November 1934 the *Flying Scotsman* set the record for reaching this long-anticipated speed. A rival company, the Great Western Railway, also laid claim to this honour for their steam locomotive *City of Truro*, which, it was argued, had travelled at just over 102 miles an hour in 1904. This was never officially verified.

Less than four years after the *Flying Scotsman*'s great achievement, *Mallard* set the world record for steam locomotion by travelling at nearly 126 miles per hour, a record that still stands to this day.

The Railway on Ice (1893)

With steam trains making headway on land, the next engineering challenge was to construct a fully functional railway on water. In Montreal, Canada, there was only one railway bridge over the St Lawrence River between the famous Victoria Jubilee Bridge and the ocean. This crossing was owned by the Grand Trunk Railroad, which charged a heavy toll to all the other railway companies for the right of traversing the water. In an effort to avoid these taxes, the South Eastern Railway Company devised its own rather inventive route across the river. Despite the obvious dangers, every winter this company used to construct a temporary railway over the ice between the districts of Hochelaga and Longueil, where the frozen river was about 2 miles wide. The rails were supported on strong pine timbers laid directly on to the surface of the St Lawrence, and used blocks of ice as ballast. This special track, which was said to be one of just two ice railroads in the world, was apparently strong enough to bear the weight of even the heaviest locomotive, brimming with goods and passengers.

As soon as spring began to thaw the frosts, the track was disassembled and the railway put in storage until the following year.

In the winter of 1923, according to the *Aberdeen Journal*, a mile-long road permitting the passage of vehicular traffic was constructed over the frozen St Lawrence River so that the neighbouring townsfolk could easily visit Montreal's carnival of winter sports. The newspaper reported on Thursday 1 February that this temporary road was well paved, and was strong enough to convey a double stream of traffic with the utmost safety. Furthermore, a line of fir trees was planted on either side of the road to form a picturesque avenue, dappled with electric lights to offer brilliant illumination at night.

The Ship Railway (1885)

With land vehicles now capable of travelling over water, the next stage in the evolution of transport was to bring water vessels on to land.

The Panama Canal is a well-known stretch of water running across the Isthmus of Panama, connecting the Pacific and the Atlantic Oceans. By 1881 work was well underway to construct this sea-level canal, which was expected to prove an excellent trade route for ships travelling from the west coast of America to the east coast. Though the plan received the support of several eminent engineers, and had by that time made progress towards completion, it had to contend with two drawbacks so serious that they threatened to put an end to the project. Firstly, the district traversed by the canal was liable to annual flooding by the swollen mountain streams, and secondly, after passing through the canal, sailing vessels would find themselves in a region of complete calm. This would mean that they would have to be towed for hundreds of miles before reaching the trade winds, thus adding enormously to the cost of the voyage.

Realising the potential impact of these setbacks, an American inventor and civil engineer Mr James Buchanan Eads (1820–1887) conceived the ingenious idea of laying down a ship railway across the Isthmus of Tehuantepec, South Mexico, instead. It was hoped that communication between the oceans would be obtained just as effectually, and with greater economy of time, by means of this railway rather than the new canal.

'The railway will be less costly to lay down than the other proposed projects,' Mr Eads argued, 'and, as compared with the canal route, will shorten the journey between New York and San Francisco by 2,900 miles, between New Orleans and the same port by 3,500 miles, and between Liverpool and San Francisco by 600 miles.'

Eads' plan for working the railway was said to be simplicity itself. A huge cradle, the framework of which was furnished with a large number of wheels, was placed in position on a pontoon. The ship entered the cradle and was then raised out of the ocean by the pontoon, and, as it left the water, was shored up for greater security. As soon as the wheels of the cradle came flush with the rails on dry land, powerful locomotives drew the vessel along the railway for hundreds of miles. When the ship reached the second ocean at the other end of the railway, the process was reversed – the cradle being run on the pontoon and, when the water was deep enough, the ship was released to float once more.

Eads obtained a concession for his inventive scheme from the Mexican Government in 1881, and wasted no time in getting to work. The surveys were completed, and the undertaking offered much promise. That was until another party made a proposal for an ordinary railway, and the government granted permission for this to be built instead.

Construction of the Panama Canal was completed in 1914, attracting 1,000 vessels in that year alone, and it remains a significant feature of world shipping.

The Land Yacht (1887)

Britain's own version of a land-based 'ship' was designed by the firm of carriage builders from Newcastle-on-Tyne. Though not quite as spectacular as Eads' ambitious concept, this little 'land yacht' was said to be extremely useful to small parties embarking on pleasure trips around the country. The horse-pulled 'yacht',

depicted in the illustration, was exhibited at Newcastle's Royal Jubilee Exhibition of 1887, and carried four companions some 300 miles all over the country in a fortnight.

The Sail Waggon (1884)

The following picture illustrates an American 'land waggon' propelled along the land by sails. The short axle of the rear wheels was pivoted by a bolt. To this axle was attached a gear wheel, into which meshed a smaller wheel, secured to the lower end of a vertical shaft. Upon the upper end of this shaft was a hand-tiller to guide the waggon. The speed was regulated by brakes on the front wheels, connected with an upright lever pivoted in the middle part of the frame, and provided at its upper end with a crosshead, so that it could be worked with the hands or feet. The vehicle was particularly suited to running on an open plain, a long road, or a sandy beach with considerable velocity.

The Flat-Packed Steamship (1884)

A Victorian land vessel of particular note is the eponymous *Le Stanley*, which was launched on the Congo River in 1884 for the use of Sir Henry Morton Stanley, the African explorer, and his team of travellers. The remarkable novelty of this jungle steamer was that it was entirely portable. It was designed to be disassembled with the greatest of ease, and the pieces were packed on the back of a waggon for transportation overland, then built up again when another stretch of water was reached. This strategy was especially useful whenever impassable obstacles in the jungle terrain prevented the ship's passage, and thus hitherto unnavigable regions of this part of the world could at last be reconnoitred by Western explorers.

This new flat-packed invention sounded terrific in theory, but considerable difficulty was met when the time came to actually haul the iron vessel for miles across the hot, craggy terrain. By the end of the expedition around one fifth of the party had been injured or totally incapacitated following a complete loss of control of the iron waggon as it careered down the stony hills of the jungle. *Science*, the illustrated American journal, reported on 1 May 1885 how fortunate it was that only two crew members had been killed during that inaugural journey.

The Paper Ship (1884)

Another seafaring vessel that sounded wholly unsuitable was launched at St Petersburg, Russia, in about 1884. This curious craft was supposedly constructed out of paper, but was much more than a mere child's toy. She was driven by steam, and was 25 feet long by 5 feet wide amidships. Her draught was remarkably small, owing to the buoyant character of her paper-made hull, and she was particularly well adapted for shallow waters.

Following the evident success of the launch, the United States Government was said to have ordered several torpedo launches whose hulls were also made of paper.

The Unsinkable Luxury Liner (1887)

The introduction of intriguing new vessels inspired other ship engineering companies to experiment with their own crafts, not just in terms of buoyancy, but also with regard to passenger comfort. With the tourism trade on the rise, the market was thrown open for cruise lining companies around the world to offer pleasure trips. In a bid to out-do the competition, ocean liners began offering every conceivable added luxury to their ships, such as silver service, deluxe cabins, and evening entertainment.

A quarter of a century before the 'unsinkable' *Titanic* embarked on her fateful maiden voyage, the construction of a splendid, lesser-known American steamer was underway at the Arrow Steamship Company's yard in Alexandria, Virginia. It was claimed that this first-class luxury liner was not only unsinkable but also fireproof, and once completed would be the fastest boat afloat. SS *Pocahontas* had been designed to cross the Atlantic within six days: a remarkable feat of engineering for that time, which promised to revolutionise ocean steam navigation.

By 1887 the architect Mr Robert M. Fryer had devoted twenty-one years of his life to the development of this wondrous steamship, and he promised that his liner would be launched before the first winter's snow of 1888 touched the ground.

The total length of the ship was 540 feet, her breadth 40 feet, and her depth 46 feet, while her draught was 25 feet 3 inches. The length of her keel, which was practically a solid mass of iron, was 510 feet, and the total weight was 710 tons. The walls and decks were constructed of steel, and the vessel contained 1,060 airtight compartments, of which 500 were below the waterline. These airtight compartments were connected with an air forcing apparatus, and filled with compressed air under a pressure greater than that of the water outside, so that in the unlikely event of a hole being made the water would be kept out. The engines, at the bottom of the vessel, were expected to generate the equivalent of 28,000 horsepower.

With regard to the fireproof arrangements, carbon dioxide gas was kept in pressurised tanks and distributed in pipes throughout the ship, so that if a fire broke out, the gas would quickly stifle it. Very little wood was used in her construction, and airtight sliding doors and shutters of steel were employed to isolate any fires.

The marble floors were covered with fine rugs, the steel walls with draperies, and the modern electric light was also fitted. Her grand saloon was 400 feet long, and decorated with ornamental ironwork and mirrors. On the main deck there were eighty-four staterooms, to which a supply of pure air and water was provided. There were also hot and cold baths in every room, and a buffet was provided at which travellers could have food at any hour of the day. In fact, the *New Albany Ledger* declared on 25 May 1887 that *Pocahontas* afforded to her passengers such luxurious comfort that cruising became a 'dream of delight'.

'Her method of construction is, beyond all question, the strongest possible,' the newspaper affirmed. 'As a whole she will be immeasurably beyond any seagoing vessel the world has ever seen.'

She was manned by 128 officers and crewmembers, and was so built that she could be readily fitted with armour in case of war.

Though no documented evidence of the ship's launch has thus far been discovered, on 30 April 1904 at Richmond, Virginia, a steamship named *Pocahontas* caught fire at the docks before finally sinking at 11 o'clock that evening. The entire city turned out to watch the blaze, which was said, according to the 1 May edition of the *New York Times*, to have been a magnificent sight. The cause of the inferno was never explained.

The Submarine (1887 & 1892)

The notion of a nautical propelling machine that would enable sailors to venture *under* rather than merely across the surface of the waves was in no way novel. Indeed, the great Renaissance master, Sig. Leonardo Da Vinci (1452–1519), left behind detailed diagrams of his own subaquatic visions. The earliest known submarine was built a century after Da Vinci's death when Mr Cornelius Drebbel (1572–1633), a Dutch inventor at the court of King James I, unveiled and tested his underwater rowing boat in the 1620s. Over the centuries that followed, many other engineers devised their own submarine vehicles that were propelled through the water by manpower. It was not until

after the development of the steam engine during the Industrial Revolution that things began to look really promising for maritime machine makers, and by the middle of the nineteenth century, following the advancement of the combustion engine, the first engine-driven submarines were at last a reality.

In 1887 a new type of submarine boat, designed by the Swedish engineer, Dr Karl Gustaf Patrik de Laval (1845–1913), was tested at Stockholm. It consisted of two boats, one on top of the other, which were joined together by long pillars. One vessel was 60 feet long and was entirely submerged, while the other, smaller one, which resembled a steamboat, remained above the water at all times. In the lower boat was a condensing steam engine and naphtha cisterns, naphtha being the fuel employed to power the submarine. The engines were worked from the upper boat, so that no member of the crew needed to be present in the lower boat when it was in motion.

Another peculiarity of the new craft was 'air lubrication'. A watertight frame surrounded the lower boat, which was filled with a supply of air. When the vessel was in motion this air was forced out into the water, thus producing air lubrication around the submerged part of the ship. Resistance between the hull and the seawater was thus reduced, thereby allowing the vehicle to run more smoothly and efficiently.

During the summer of 1892 another submarine, invented by a gentleman named Mr Baker, was successfully tried on the Detroit River, United States, and could travel up to 10 miles an hour. As shown in the sketch it consisted of a capsule-shaped hull, built of stout oak planks, and capable of withstanding the pressure of the surrounding water at a depth of 80 or 100 feet. It was fitted with two screw-propellers, S, one on each side, and the rudder, R, sat close to the hull. The conning tower on the top is shown at T.

When the boat rose to the surface an internal steam engine was connected to the propellers, but when she was submerged, a feat achieved by pumping water into her tanks, the steam engine was discarded, and an electric motor, fed by charged accumulators, was brought into gear with the screws.

The interior of the craft was lit by electricity, and could accommodate several men. Only two were required to manipulate the vessel, and the hull, which was 40 feet long, contained sufficient oxygen to allow their being underwater for about three hours.

The Telescopic Submarine (1887)

The *Nautilus* was another electrically powered submarine, intended to be operated as both a torpedo and search boat. With the help of a team of inventors, the British engineer Mr Andrew Campbell devised this innovative machine in 1886.

The vessel, which was named after Jules Verne's fictional submarine, was raised and lowered in the water by expanding or contracting in bulk, or in other words, by altering its power of flotation. This was achieved by sending in or out eight telescopic chambers that projected from the vessel, four on each side. When these chambers were wound in, by an increase of displacement the vessel became less buoyant and it instantly sank out of sight. By a reversal of the process it floated back up to the surface again.

The cigar-shaped boat was propelled by electricity stored in accumulators, and lighted by the same means. A supply of compressed air was provided for the men on board, enough to last for twenty-four hours. The crew entered the vessel through a manhole at the top of the hull, which was sealed tightly so that neither water nor air could enter or escape until the cover was removed again.

On Saturday 27 November 1886 several dignitaries were invited on board as part of a trial run at Tilbury Docks, Essex. Their number included Admiral Lord Charles Beresford M.P. (1846–1919) and the illustrious warship designer Sir William Henry White (1845–1913). At first the *Nautilus* seemed to run to satisfaction, with the *Portsmouth Evening News* describing on 29 November how the vessel instantly disappeared beneath the murky water once the guests and crewmembers were on board, before bobbing back up again 'with the buoyancy of a cork'.

Despite the apparent success of this maiden voyage, there was an obvious flaw to the design: once the batteries had been used – and there was no facility for recharging them – the submarine could not be moved. After several trips through the water, the submarine went down and did not reappear again for around a quarter of an hour, by which time the spectators began to worry about the safety of all on board. It transpired, as reported on 3 March 1888 by the *Manchester Courier and Lancashire General Advertiser*, that the vehicle had run out of power and became stuck in the mud on the bottom of the Thames. Although everyone was eventually able to get back to the surface again safely, the opportunity of another ride was swiftly refused.

The Balloon Railway (1885)

By the end of the century, anything seemed possible. Submarines were carrying passengers deep below the waves, steam trains were rattling across the landscape, and the telpher line had given rise to an early form of aerial transportation. Since the ancient days of Icarus, mankind has been driven by an overwhelming desire to defy gravity with the help of mechanical wings. After generations of dreaming, had the time finally come for man to invent a vessel that could sail across the skies? A small group of bold Victorian engineers believed they now possessed the tools and expertise to achieve the miracle of flight, and they began their avian mission by modifying the tried and tested design of the locomotive.

In 1885 a vertical railway, where the cars were to be carried upwards by the levitation force of a great balloon, was planned for construction on the Gaisberg, near Salzburg, in order that passengers may be raised to the summit of the mountain to admire the spectacular views on offer. The vehicle was to have grooved wheels on one side of its car, and would ascend and descend a perpendicular line of rails. Such ambitious plans, however, were not immediately realised, although funicular, or cliff, railways, which utilise cables rather than balloons, can still be found travelling up and down steep hillsides across the world.

The Clockwork Bird (1884)

It was the French who paid particular attention to the subject of aerial locomotion during the nineteenth century, and the French Academy of Sciences in Paris awarded three prizes to pioneers in this domain. Monsieur Gaston Tissandier (1843–1899), the aviator who made successful trips through the air in a balloon propelled by electricity, was one of the prize winners. Another was Monsieur Victor Tatin (1843–1913), who designed the membrane propeller of Monsieur Tissandier's electrical balloon. Tatin's prize was, however, obtained for his invention of his ingenious Oiseau Mechanique – an artificial bird that mechanically mimicked the flight of a living bird by strokes of the wing, its power derived from rubber bands that were wound by clockwork inside the body.

The Dirigible, La France *(1884)*

It was testament to France's commendable determination that the world's first controlled airship flight to return to its original starting point took place in French skies. On 9 August 1884 a momentous balloon ascent was made from the French munici-pality of Meudon by two pioneering government aeronauts, captains Charles Renard (1847–1905) and Arthur Constantin Krebs (1850–1935).

Their airship, named *La France*, was driven by a screw propeller and steered by a rudder shaped like a large square sail. captains Renard and Krebs had been engaged for some years with experiments on ballooning at the government establishment in Meudon, and had built the dirigible out of light, strong silk. It was cigar-shaped and pointed at both ends, measuring 197 feet in length and 39 feet in diameter. A light netting covered the balloon, and supported a platform 131 feet long by 10 feet broad. This was the basis for the car which carried the passengers and freight. The propeller, which consisted of a screw made from a light wooden framework and airtight cloth, was fixed to the front of the platform. It was driven by a dynamo, actuated by a current from a set of stored accumulators.

Renard and Krebs ascended in *La France* together, the former gentleman having charge of the propelling gear, the latter of the rudder. On being liberated from the ground, the aerostat rose quickly to a height of 180 feet, and then, the propeller being in full play, started off for Villebon, a town 7 miles distant. The day was calm, but there was a breeze blowing against the balloon at the rate of 18 feet per second.

On arriving over Villebon the aerial craft was steered gradually around, performing a perfect U-turn, and started back towards Meudon, where it landed safely some forty minutes after it had set out.

In 1886 the Académie des Sciences awarded their third and final prize to captains Renard and Krebs for their ground-breaking contribution to aeronautics.

The Flying Machine (1892)

La France was exhibited at l'Exposition Universelle in Paris for six months where it became an inspiration to many aspiring airship captains and engineers. One such man was Monsieur Gustave Trouvé, the distinguished inventor, whose own experimental flying machine, or *'aviateur'*, was introduced to the academy in about 1892, and was considered both highly ingenious and original.

Monsieur Trouvé reasoned that none of the existing motors of the nineteenth century – whether powered by steam, electricity

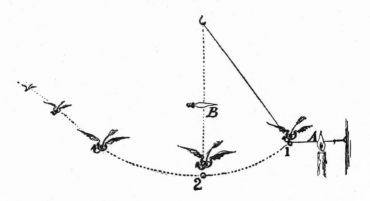

or compressed air – were capable of successfully propelling a vehicle through the air. In order to devise a new type of motor that could rise to this challenge, he took up the principle of the Bourdon tube, which was employed as a manometer or pressure gauge. This kind of tube was constructed in the shape of a horseshoe, and when filled with gas, the points of the tube approached towards or receded from each other accordingly, as the pressure of the gas fell and rose.

Trouvé theorised that he could transform this kind of tube into a motor by filling it with an inflammable mixture of hydrogen and air, and exploding the mixture with cartridges that were fixed on a revolving barrel. Upon each explosion the pressure of the gas in the tube would fall, and the points of the tube would approach. On refilling the tube with gas between the explosions, the points would recede. In this way a reciprocating motion of the points would be obtainable as long as the supply of hydrogen and the exploding cartridges lasted. To utilise this motion, Trouvé fastened an expanded artificial wing to each point of the tube, which would rise and fall as the points moved, thus beating the air like a bird's wings.

His fanciful *aviateur* is illustrated on the previous page, where the wings, A B, are seen attached to the points of the horseshoe-shaped Bourdon tube. D was a reservoir of compressed hydrogen, and C was the body of a small aeroplane, indicated by dotted lines.

The figure above represents a small model of the *aviateur* starting on its flight. The model is shown suspended by a cord from a bracket and held in position at point one until the restraining tie, A, was burnt through by the flame of a candle. It then swung to position two, where a second flame, B, burnt through the sustaining cord. By this time the series of twelve cartridges would begin to explode at regular intervals, causing the Bourdon tube to react. It was at this stage that the *aviateur* was, in theory, able to support itself in the air by the action of its wings, and even mount upwards in its flight. The dozen cartridges were said to be able to propel the vehicle from 75 to 80 metres above the earth.

Despite his best efforts, Trouvé's machine never got off the ground, though other nineteenth-century inventors and engineers continued doggedly in their hunt for a fully functional flying machine.

The Aerodrome (1893)

The image overleaf represents a working model of the piloted flying machine, or 'aerodrome', which Professor Samuel Pierpont Langley (1834–1906), an aviation pioneer of the Smithsonian Institution, Washington, had invented by 1893, after nearly two decades of steely devotion to the study of mechanical flight. His winged vehicle, unveiled an entire decade before Messrs Orville (1871–1948) and Wilbur (1867–1912) Wright had entered the arena, was intended to carry passengers. The vehicle, it was noted, bore some resemblance to a dragonfly, though some christened it 'the flying fish', the body being shaped like a 15-foot mackerel.

The backbone of the machine was constructed from a tube of aluminium steel, and ribs of the same metal were employed to give the framework sufficient rigidity. The double oscillating engines were located in the head of the 'fish', while the four copper boilers occupied the middle, and instead of water they were fed with a volatile hydrocarbon, which vaporised at a low temperature. The fuel was refined gasoline, stored in a tank at the tail of the vehicle. The body was insulated with asbestos to prevent loss of heat, and the wings, or aeroplanes, consisted

of aluminium steel frames covered with China silk. The pair in front were 42 inches wide and 40 feet from tip to tip, and could be adjusted at different angles. The propeller consisted of twin screws capable of being turned to varying degrees to allow steering. A cross-piece, or tubular mast, ran through the body of the machine, and stays of aluminium wire tied the whole apparatus together.

At first, the vehicle's constructors fought to preserve the secrecy surrounding the flying machine, but sufficient details were leaked to make the impending result of Professor Langley's experiment a matter of great interest in scientific circles. Initial trials of the aerodrome were conducted using unmanned models, and it wasn't until 1903 when the first full-size flights were made. According to a story in the *Derby Daily Telegraph*, published on Thursday 8 October, Langley's flying machine was a disastrous failure. Despite the American Government investing £15,000 in the invention, the aerodrome crash-landed into the Potomac River, Virginia, moments after its catapulted take-off, and was completely wrecked. The pilot escaped without injury, although the whole folly was said to have crushed Professor Langley's spirit, and popular belief held that he died three years later of a broken heart.

In 1914, according to an *Evening Telegraph* article dated 1 June, Langley's obsolete aircraft was tested one final time and found, in fact, to be a practicable aeroplane. The mistake Langley made was catapulting the machine from a height rather than propelling it along the water's surface and allowing it to rise steadily into the air under its own steam. Needless to say, it was the legendary Wright brothers who were credited with inventing and flying the first successful motor-driven aeroplane in 1903.

The Giraffe Hotel (1884)

Since the invention of passenger aeroplanes, the world's tourism industry has blossomed into one of the largest service industries in history. The foundations of this modern money-making business date back to the nineteenth century, when more efficient forms of transport first became available, offering the wealthier classes the opportunity of paying visits to picturesque places they had hitherto only read about within the pages of guide books. Mr Thomas Cook (1808–1892), the founder of commercial tourism, began offering all-inclusive tours in 1841, and his success inspired other travel agencies to be established. As the industry continued to flourish, hotels and guesthouses all over the world began to materialise, opening their doors to the weary traveller.

In 1884 a curious structure in the form of a gigantic giraffe-shaped hotel was built at Sheepshead Bay, Rhode Island, directly over the Jerome Hotel, to which it formed an annexe. The structure was said to be 280 feet high, and stood on four legs, three of which were used as stairways, the fourth as an elevator. A large seafood restaurant, 40 feet by 60 feet, occupied the body of the animal, and a traditional clam-bake dining saloon was fitted up in the head. As a finishing touch, two electric lights beamed out from the eyes of the hotel, casting their rays far across the sea.

The Hotel Boot Label (1884)

The accompanying engraving illustrates a Victorian convenience that was designed to ensure that travellers who stayed in hotels or private residences received a wake-up call at the desired time in the morning. The old custom was for the guest to leave their footwear with the proprietor, and for 'boots', as they were termed, to chalk a number on the soles indicating the hour at which the boarder wished to be roused the next morning. Unfortunately, this plan sometimes led to inconvenient mistakes; ergo boot labels were brought out. These were hung in appropriate places, such as the coffee rooms and bedrooms of a hotel, so that when the guest removed his footwear he was able to write the desired hour of his morning call on the label and affix it to his own boots prior to handing them over. One type of label showed a clock face, which could be set to the required hour, and another, shown in the illustration, allowed provision for ordering a bath or breakfast. This convenient system permitted the traveller of a peaceful night's sleep, free from any worry about a late emergence. He would wake fresh in the morning, put on his boots and be on his way at the start of a new day ... and very much a new dawn.

INDEX OF INVENTIONS

A Marvel in Clocks 76
Accident Ambulance 187
Acme Thunderer Whistle 248
Adjustable Braces 71
Aeolia Harp 60
Aerodrome 283
Airbrush 82
Allan Glen Revolver 86
Ambulance Steamer 189
Ammoniaphone 58
Asbestos Hat Linings 70
Asylum Gong 133
Automatic Cotton-Picker 106
Automatic Egg Cooker 13
Automatic Piggy Bank 120
Automatic Postman 30
Automatic Telephone
    Exchange 239
Automatic Well Cleaner 24
Balloon Railway 278
Bicycle Ambulance 188
Bicycle Railway 252
Body Lightning Rod 170
Brachionigraph 217
Brake for Perambulators 37
Bronchitis Tube and Kettle
    195
Bucket Fire Escape 185
Businessman's Dinner
    Table 92
Buttonhole Flower Holder 75
Cab Communicator 259
Cable Tramway 260
Camel-Drawn Monorail 259
Camera Obscura in Coastal
    Defence 157
Candle Motor 49
Carpet-Sweeping Machine 17
Casting Metals on Lace 79
Champagne-Bottle Drain 223
Chanticleer Preventer 43
Checking-In Clock 95

City Fire Alarms 173
Clockwork Bird 279
Coloured Photographs on
    Glass 161
Combination Brush 24
Combination Spray Washer
    19
Combined Couch and Settee
    28
Contractible Perambulator 36
Convertible School Furniture
    88
Cooking with Gas 10
Copying Apparatus 91
Crime-Fighting Streetlamp
    145
Crystal Chair 77
Curious Needles 75
Darkness Photographs 149
Deep-Sea Photography 152
Diminutive Dictionary 84
Dirigible, La France 279
Disinfecting Lamp 197
Disinfecting Oven 199
Diver's Helmet 204
Domestic Ozoniser 195
Earthquake Recorder 215
Egg-Boiling Spoon 15
Electric Burglar Alarm 136
Electric Chicken-Hatcher 139
Electric Gun 126
Electric Jewels 73
Electric Lamp Clock 132
Electric Lifeboat 128
Electric Light Spectacles 142
Electric Parcel Post 135
Electric Power for Paris 125
Electric Underground Parcel
    Exchange 133
Electrical Light Effects 143
Electrical Ore Finder 213
Electrical Trumpet 137

Electrical Utensils 137
Electrical Wedding 140
Electricity from the Wind 229
Electro-Photographic Thief
    Detector 136
Exploding Scarecrow 101
Fanned Rocking Chair 27
Ferris Wheel 45
Fire Exits 179
Firedamp Detector 202
Firewood Machine 106
Flat-Packed Steamship 272
Floating Hospital 190
Flying Machine 281
Folding Music Stand 64
Foot Dustpan 18
Foot-Warmer for Trains 264
Forth Bridge 111
Fruit Gatherer 102
Garden Waterer and Roller 54
Gardening Tool 56
Gas Flat Iron 21
Gelatine Hygrometer 219
Geodoscope 85
Geological Piano, The 61
Ghost Train 263
Giraffe Hotel 285
Glass Floor 117
Glass House 9
Glass Paper and Pulleys 118
Glass Sewer 117
Gold Finder 214
Grain Dryer 99
Gravimotor 47
Great Equatorial Parisian
    Telescope 211
Greenhouse Heat Alarm 172
Grinnell Fire Sprinkler 178
Gymnastic Treadmill 48
Hand Fire Extinctor 176
Hand-Grenade Fire
    Extinguisher 174

Hands-Free Home Telephone 233
Health Clothing 68
Hearing Test 192
Heat-Sensitive Paint 226
Heating Railway Cars by Gas 264
Heliochromoscope 158
Hinged Lamp Post 145
Hippometer 256
Hoeschotype 81
Homacoustic Speaking Tube 232
Hospital Photography 150
Hotel Boot Label 286
Household Blowtorch 23
House-Warming Apparatus 26
Hydrophone Torpedo Tell-Tale 129
Hydrophone 221
Hypnoscope 217
Ice Harvester 97
Iceberg Detector 222
Illuminated Rose 74
Illuminated Train 132
Incandescent Streetlamp 144
Indicating Cooking Skewer 12
Indicating Door Bolt 93
Influence Machine 122
Infusing Teapot 15
Ironing Machine 22
Koniscope 193
Ladder Fire Escape 183
Land Yacht 270
Largest China Vase in the World 77
Letterbox Annunciator 32
Life Preserver at Sea 202
Life-Saving Pillow 205
Lightning Guard 123
Long-Distance Photography 153
Luminous Paint 226
Magazine Gun 127
Maxim Gun 127
Mean-Time Sundial 224
Mechanical Telephone 235
Menthol Inhaler 193
Microphone in Medicine 191
Milk Receiver 33
Miniature Sun 131
Moveable House 7
Moving Photographs 166
Music Writer 63
Niagara Falls Hydropower 114

Niagara River Cantilever Bridge 113
Office Indicator 93
100 Miles an Hour by Rail 267
Opera-Glass Camera 155
Optical Performance 168
Osmium Telephone 237
Oven Heat Indicator 11
Paper Ship 273
Parcel Post Balance 30
Pencil Suspender 72
Phonoscope 165
Photographing the Walk 162
Photographs of Lightning 150
Picture Telegraph 245
Pictures in Stone 80
Pike's Peak Railway 261
Pocket Ambulance 186
Pocket Anemometer 212
Pocket Rack for Coats and Hats 39
Portable Aluminium Army Canteen 110
Portable Columbia Typewriter 90
Portable Greenhouse 8
Portable Railway Bridges 110
Portable Saddle Fire Escape 181
Postage Stamp Seller 118
Potato Digger and Picker 102
Prescribing by Cable 243
Prison Puzzle 50
Projecting Microscopic Views 210
Quadrant Tricycle 251
Railway on Ice 268
Reading Chair 65
Reflecting Oven 10
Remarkable Uses of the Telephone 238
Road and River Cycle 253
Rowing Machine 47
Rug Machine 108
Sail Waggon 271
Sandwich Indicator 53
Scoto-Irish Telephone 243
Screw Lift for the Eiffel Tower 44
Sea Telephony 242
Self-Cleaning Garden Rake 57
Sewer Gas Cremator 116
Sheep-Shearing Machine 107
Ship Railway 268

Shop Weigher 119
Signal Light of the House of Commons 147
Small Writing 84
Soil Heating 53
Solar Battery 228
Solar Panels 227
Spark Photography 162
Spectacles for Horses 257
Steam Man 103
Steam Tree-Feller 105
Stenotelegraph 247
Straw Shoes 70
Straw Villa 9
Submarine Electric Lamp 131
Submarine 275
Sun Motor 227
Sunshades for Soldiers in the Desert 171
Sun-Wound Clock 225
Swan's Electric Safety Lamp 200
Tea Dryer 100
Teaette 17
Telautograph 244
Telegraphing 7,000 Miles 243
Telephotography 246
Telescopic Submarine 277
Tell-Tale Milk Jug 35
Telpher Line 260
Time by Nightlight 41
Tintometer 212
Tornado Photographed 152
Tower Bridge of London 112
Travel Holdall 265
Travelling Office Chair 89
Travelling Platform 45
Travelling Telegraph 241
Treadle Driving 255
Treasure Safe 208
Tricycle Chair 250
Trunk Lifebuoy 207
Umbrella Clock 40
Underwater Spyglass 51
Unicycle Sleigh 254
Unsinkable Luxury Liner 273
Uses for Asbestos 109
Voting by Electricity 141
Vulcan Water Sprinkler 176
Watchman's Tell-Tale 96
Water Bell, The 52
Whispering Machine 67
Whitby Lifebuoy 206
Wireless Telegraphy 240
Xylophone 62